Once Upon a Tar Creek

Mining for Voices

by

Maryann Hurtt

Turning Plow Press

For Bob-by,

Once Upon a Tar Creek

Mining for Voices

So happy to be connected again!

by

Maryann Hurtt

Maryann-Wendy
2022

Cover Image: Vaughn Wascovich
Book Design: Rowan Kehn

Turning Plow Press

ISBN: 978-1-7355762-2-0

For the Tar Creek Monster

who was the initial inspiration for this book

and for the monsters

in all our lives

perhaps all the dragons
of our lives are princesses
who are only waiting to see
us act, just once, with beauty
and courage, perhaps every
terror is, in its deepest essence,
something that needs our
recognition or help

Rainer Maria Rilke

Author's Note

Who talks is always a struggle when writing. Persona poems can be especially difficult. As I listened to the stories coming out of Tar Creek, both people and what we normally think of as inanimate became real to me. The voices of a burning Ku Klux Klan cross, a toxic orange-water creek, a long-gone pillar of the community, soon felt like hearing next door neighbors. Yes, the poems in *Once Upon a Tar Creek: Mining for Voices* are based, at least loosely, on actual events, people, places, and creatures, but imagination, too, became part of the reality.

As I listened to and researched dozens of stories I struggled with questions of cultural appropriation. With all the respect I can muster, I now place the stories in your hands believing in the truth of all our lives. Maybe the more we know about what we think of as "other" the more we will realize we're hitched to everything and everyone in the universe.

Any mistakes are my own.

What is history without imagination?

Herodotus (484-425 BCE)

Greek Historian

Folks think there are five senses. Nah, six. Gotta have imagination. That's when you start to understand something.

Arnold Richardson (1906-2008)

Honorary Modoc Member

Oklahoma Centenarian

When we try to pick out something by itself, we find it hitched to everything else in the universe.

John Muir (1838-1914)

Scottish-American Author and Naturalist

"I would ask you to remember only this one thing," said Badger. "The stories people tell have a way of taking care of them. And learn to give them away where they are needed. Sometimes a person needs a story more than food to stay alive. This is why we put stories in each other's memory. This is how people care for themselves."

Barry Lopez (1945-2020)

from *Crow and Weasel*

Table of Contents

"Morning on the Creek" by American artist Charles Banks Wilson (1918-2013). He grew up swimming and fishing on Tar Creek and completed this lithograph in 1967. Permission given by the Gilcrease Museum.

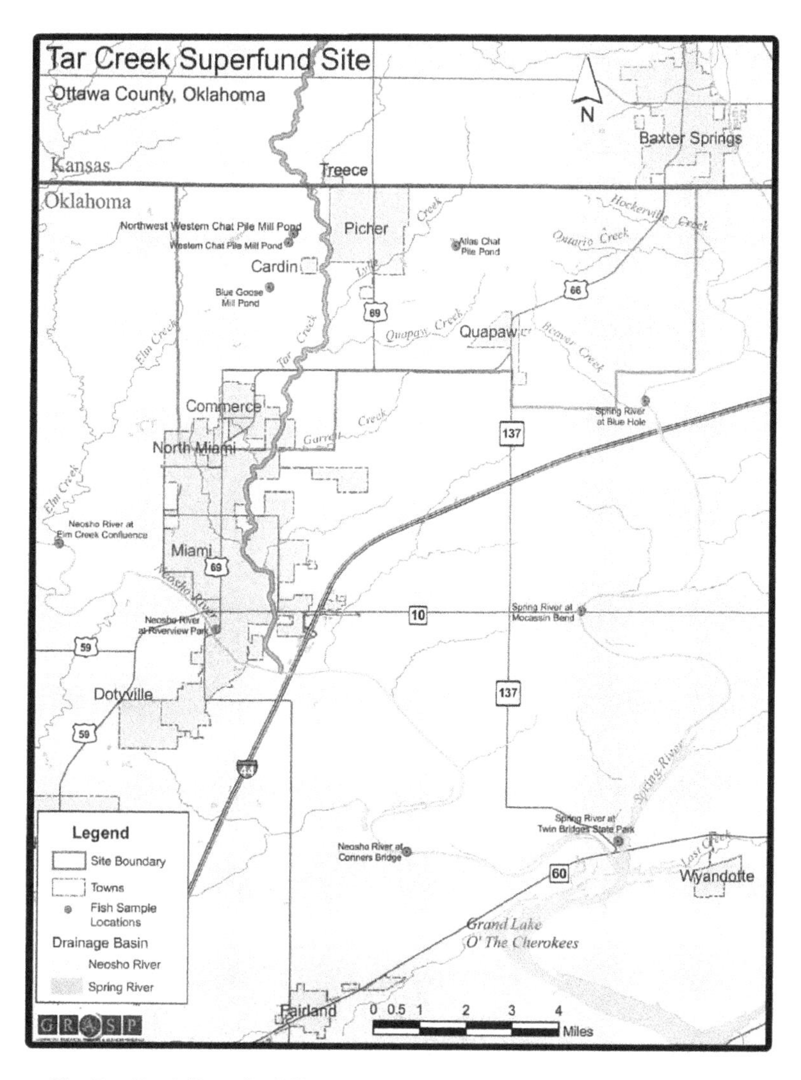

The Tar Creek Superfund Site. Courtesy of the LEAD Agency—Rebecca Jim.

Introduction

How My Heart Learned to Bleed Orange

Lead and zinc lay beneath the earth's surface and mining thrived during times of world wars. Some people became rich, but many others suffered and endured acts of injustice. Here are the stories whose voices need to be remembered. We are still seeking justice. Our story is not over. Wakonta is with us.

Grace Goodeagle Ogaxpa/Potawatomi,
a Redeagle and Quapaw member.
Quapaw Tribal Chairman, 1994-1996

Sometimes I believe I divide my life before Hat Creek and after Hat Creek.

When I was four years old, my parents took me to Hat Creek in northern California where I caught my first fish—an eleven-inch rainbow trout. I remember the smell of the water, the marshy weeds, the alien odor of a creature so different from myself. Most of all, I remember the sense of being home. The water that ran through Hat Creek was cold and clear. I could drink the water. It was a place I wanted to and did go back to many times. Hat Creek taught me fierce love—the kind that makes me want to protect and fight for what is precious and good in life.

Over and over again, I am drawn to water. It's where I

find comfort both in good and bad times.

Orange Water

Fifty years after that first fishing trip, I made my way to a new creek—Tar Creek on the Kansas-Oklahoma border. My mother grew up in the area where my grandfather worked in the lead and zinc mines. Oklahoma was still Indian Territory when my grandmother was young and she and my great-grandmother both worked at the Quapaw Indian Agency. A family reunion brought me to Ottawa County.

My early morning ritual usually involves hiking, running, walking, and finding the nearest water. That morning in 1990, I found a cross-country track next to Northeastern Oklahoma Agricultural & Mechanical College in Miami. The track ran next to water and I had my first sighting of a nightmare called Tar Creek. Nothing could have prepared me to see such horror. I thought vandals had purposely sprayed neon orange paint up and down the creek banks and in the water itself.

My first week at Tar Creek went by quickly. I couldn't stop asking questions. "How does a creek turn orange? What about the lead mines? When did the mining start? Are the mines still running? Who lived on the land where the lead was found? Did the tribes that run casinos all over the county always live here? What are the tall white hills in the middle of prairie land?"

I talked to ranchers at the feed mill, the bookstore owner, a newspaper editor, casino players, 2nd and 3nd cousins who never left Oklahoma. My mother and I drove out to Picher where most of the mining had taken place since the early 1900s. An old miner took us through the Picher Mining Museum. The building had been the headquarters of the Tri-State Ore Producers Association and later their mining health department. Hundreds of miners' pictures lined the walls. We found my grandfather's photo and his health department card—"William H. Cantwell, age 38, 1939, two days in ground" (referring to starting work in the mines two days earlier).

The mines closed back in the early 1970s but I was told the "white hills" are actually 200 feet tall piles of chat left over lead and zinc tailings. The orange water was acid mine water from the abandoned mines seeping underground into Tar Creek. I read a *Los Angeles Times* article by Richard E. Meyer, calling the mines "a 10-billion-gallon vat of subterranean poison." The United States Environmental Protection Agency declared Tar Creek a superfund site and one of the worst hazardous waste sites in the nation.

The more I studied Tar Creek, the more I was drawn to its history. Water, in all its shapes and forms, has always captivated me. To see such defilement in what I loved both terrified and outraged me. The more I learned, the more questions I had. When I returned to Wisconsin, I read everything I could about Tar Creek. And I knew I had to go back.

Back to Tar Creek: Meeting a Monster

A year after the family reunion, my mother and I began annual trips to Oklahoma.

We stayed in a tiny motel room and got out early to explore family cemeteries, old school houses, and back roads. I drove the rental car, she called out directions in her long "pee" skirt worn specifically for side-of-the-road pit stops. She amazed me with her memory of school bus routes and the homes of kin long gone. And of course, always back to Tar Creek.

My mother told me her own mining stories. She remembered her father stretching out his arm and how she and her brothers and sisters could swing from it. Picking and shoveling lead all day built incredible muscles but destroyed lungs. I learned about "miners con." (what the miners called silicosis) which often led to tuberculosis. In the 1930s, Ottawa County, Oklahoma had the highest TB mortality rate in the United States. My grandfather ended up in the Oregon state TB sanitarium after becoming part of the Okie migration at the end of the Dust Bowl era—his dust was *chat*, left over mine tailings. Even at eighty years old, my mother still looked scared when she talked about mining accidents and a little girl's fear that her father would not come home some night.

One blue sky-nothing-could-be-bad morning, I left the motel room and went down to Tar Creek. I watched the water for any strange doings and imagining that the pumpkin-colored creek held. Two turtles placidly cruised the water; watching them I wondered about possible extra appendages, frail shells, and other weird genetic mutations.

A rustle in the water caught my attention. A long-necked almost snake-like creature stuck its feathery body out from the creek's bottom. We studied each other with no fear, just absolute fascination. It seemed we were waiting for this moment. The creek went still, birds stopped singing, my scrutiny was focused on the alien creature. In all my years wandering creeks, rivers, lakes, I never encountered anything or anybody like this. Was this creature the result of an environment so toxic it would produce extreme changes?

We continued to stare at each other as if we wanted a common language. But our connection was beyond language, and in some way I knew the creature intimately. We didn't need words. The water connected us. Then as quietly as it first appeared, the snake-bird slipped back into the creek and was gone. But not before I made a vow to this strange creature, or I just made a vow to myself, never to forget the voices rising out of Tar Creek.

I go back to Oklahoma on my own now. My mother died several years ago but she taught me well how to navigate back roads and how to listen to stories. In 2011, I returned.

Old Stories, New Stories

My plane landed late at night in Springfield, Missouri where I rented a car for the two-hour drive. Country music stations kept me awake as I drove between big semis heading west on the interstate. I watched for the exit that would take me to the cabin I rented on Spring River. The roads got darker and darker and went from pavement to gravel to dirt by the time I finally arrived.

Tar Creek flows into the Neosho River and the Neosho River and Spring River meet a few miles from where I would be staying. The cabin fit my needs perfectly—from it I could find the people, places, and stories I wanted to learn.

I had made a "who, what, when, where, why, how" list and rehearsed how to introduce myself. "Hello, my name is Maryann Hurtt, my family was from out near Wyandotte and my grandpa worked in the mines. I'm here visiting and trying to learn some more about the mines and the history around here. Any chance I could spend some time with you? Several folks told me you were a good person to talk to." I wanted to help preserve the stories, but I didn't want to scare people off.

It seemed especially important to talk to old miners since so many were dying off, both from old age and many times prematurely from “miners' con.” I wanted to know about a man named McNaughton, the first white man to get a lease on Indian land to mine back in the 1890s. The folks at Dobson Museum in Miami, Oklahoma gave me the names of old Quapaw women they thought would help me. On a previous visit, I had met the Picher funeral home owner who had at one time both a funeral and an ambulance service, and I wanted to hear more.

I started with Paul and Wanda Thomas, the owners of the old funeral home. When I called, Wanda told me to come on over. She and her husband were both ninety years old and he slept through most of my visit. But Wanda, sitting alone much of the day, was eager to tell a wealth of stories. We sat in their apartment back of the new funeral home their son had built. I recognized the plush furniture and exotic vases that filled the old funeral home I had visited a few years earlier. She remembered middle-of-the-night mining accident calls, how to fix a body when it's hardly recognizable. Before I left, she told me about Bill, their friend and an old miner, and gave me his telephone number.

Back at my cabin, I called Bill Crawford. His daughter answered and yelled over to her dad, "Daddy, is it okay if this lady visits? She wants to know about the mines." He agreed and she gave me directions to their place.

The next morning I sat in Bill's workshop where old mining equipment filled the walls and pathways and listened to stories. "I couldn't bear to see it all junked when the mines closed, probably can use the stuff for something, sometime. Anyway, I like the stuff, it helps me remember." He showed me his garden, including footlong okra, and fed me okra soup. Before I left, he told me about the widow living at the old McNaughton homestead and how to get there.

Not far from Bill's home is the Ankenman Cattle Ranch, once the McNaughton homestead. Sandy Ankenman is as fascinated with history as I am. She walked me through the three and half story barn McNaughton built back in 1893. The barn had been used at various times as a jail, post office, and stage coach terminal. Sandy made me a copy of the original lease McNaughton received to be the first white man allowed to start mining on Indian land.

Ardina Moore was next on my list of contacts. She is one of Quapaw tribe's last native speakers, a respected elder, and has taught Quapaw history at the local college and Quapaw language at the tribal center. She readily shared Quapaw history and memories of her grandpa, Victor Griffin, a traditional Quapaw chief. I asked about Robert Thompson, a Quapaw man committed to Hiawatha Insane Asylum I had read about in a 1924 book, *Oklahoma's Poor Rich Indians: An Orgy of Graft and Exploitation of the Five Civilized Tribes*. She questioned why I would want to write about him. As we talked, she remembered she had heard he had stopped talking when he came back to Oklahoma.

I also knew I still needed to meet the people at the **L**ocal **E**nvironmental **A**ction **D**emanded—**LEAD** Agency**.** LEAD Agency is a grass roots environmental group that has worked for years to educate and advocate for Tar Creek.

Rebecca Jim and the LEAD staff were welcoming. Rebecca, a tiny Cherokee woman, spoke with passion that outsized her small stature. The outrage she feels about Tar Creek filled the room and wherever she goes. The old house they rent is stacked with documents, pictures, newspaper clippings, and scientific studies. We talked at length and I read them some of my Tar Creek writing. I received an invitation to read at their Tar Creek Environmental Conference.

Their commitment was contagious and I knew I had found kindred souls.

All week I followed threads, made more connections, spent time at the local college library, met more people who introduced me to more people and stories. I walked down Main Street in Miami where an old man told me about seeing the leftover embers after a Ku Klux Klan cross burning. Sitting in old cemeteries, I listened hard for ghosts who still had stories to tell. Late at night, I returned to my cabin by the river and set down lists of the voices—old Quapaw women, chat piles, peyote, Ku Klux Klan crosses, miners, school children with too high lead levels, Chief Joseph, Indian school runaways, a town too toxic and dangerous to live in.

Before Leaving

My last night of the trip, I went to Braum's for black walnut ice cream. I got it to go and headed for Tar Creek. I needed to say "goodbye" at least for now, to this creek that had become such a part of me. I walked down to the water, found a big rock, and ate my ice cream. I scanned the creek hoping for another meeting with the strange creature that set me on this course. I didn't see the snake-bird but I felt his presence in the water and the rocks and the wind changing to fall. This landscape that I had learned to love spoke to me in ways I was just beginning to hear. When I got back to my car, my socks were stained orange, and would stay that way after many washings.

It's strange how we affect each other—chance encounters that change the way we see the world and make us care in ways we didn't know were possible. Maybe my heart bleeds orange now and makes me pick up this pen and share these stories before they are forgotten.

The water, the creatures, the miners, the miners' children, the minerals, the Indians, all of us, you and me, we're in this together and we are all downstream. To know this in our deepest hearts may be our one great hope in a crazy universe. A fierce, ferocious love, the kind that gives our children's children the chance to love the earth and all its wondrous, watery ways.

The Tar Creek Monster Speaks
May 29, 1994
Miami, Oklahoma

I didn't always look this way
at one time
my feathers ruffled sunlight
made the whole world feel brand new

in those days
skinny dipping boys
belly flopped, dog paddled
down this creek
then cast lines
for my brother fish
supper for whole families
back then in Depression time

now I lay low
don't show my face
know what a fright
I really am
some kind of freak
half bird, half snake
some kind
of hybrid

nobody wants to see

but one dawn
the sun too sweet
to hide from

I chance my way
back to Tar Creek
don't expect nobody
this just start of light
blooming
into full blown day

I feel the heat
let go my dread
look up and see a woman
staring straight on
no fear either
appears we had this all planned

6:05 AM
Tar Creek Bridge
just east of NEO College
May 29, 1994

we take measure of each other

know right as rain

this is who

is ready

to hear the stories

Back in 1994, I was walking past Tar Creek in Ottawa County, Oklahoma and stared into its orange water. I wondered what "a 10 billion gallon vat of subterranean poison" would produce—a place that some say is the biggest environmental disaster no one has heard of. Out of the water, a creature unlike anything I have ever encountered raised its snaky feathered head. I now know there is a snake-bird called an Anhinga—they normally don't live in these parts. Tar Creek was the site of some of the biggest lead and zinc mining until the late 1960s-early 70s.

See Meyer (1993) for one of the first national reports about Tar Creek.

I Was That Woman

October 23, 2016
Elkhart Lake, Wisconsin

who so many years ago
stood staring
deep into water
and met a creature unlike
anything or anybody
I have or I'm sure ever will have
met
but this is what you need
to know
how on that day
looking straight on at a feathered
scaly, long-necked, beaked
creature studying a pale white
crinkle-eyed older woman
a pact was made
how in some mysterious way
I got entrusted
really the honor
to listen, know, and record
the stories
coming out of Tar Creek

The Tar Creek Monster and I made a pact and in some kind of mystifying way, the stories just keep coming out of Tar Creek. I made a promise, both to myself and to this mysterious creature, to know and record the stories.

Military Road Still Feels the Feet
April 2, 2015
Oklahoma-Kansas state line

their steps still thump

in my ears

you, too, if you listen

close enough

will hear the soldiers

cattle

beat down Quapaw

all looking for a way

in or out

on hot sweaty days

and the sky turns yellow

or when the wind comes down

from the north

and mamas trying to protect

their little ones

look to the side of me

see the lumps of those

who didn't make it

I've seen it all
Quantrill, bushwackers
jayhawkers, miners
rebel boys, yankee boys
exiled Indians
looking for safety
revenge
maybe just home
and some kind of rest

The Military Road (originating at Fort Snelling in St Paul, Minnesota and eventually leading to Fort Jesup, Louisiana and the Gulf of Mexico) enters Indian Territory-Oklahoma up near Baxter Springs and was used by pioneers, soldiers—Yankee and Confederate, cavalry, Indians, miners, prisoners of war. In Oklahoma, the Military Road is known as the Texas Road. Some places you can still see the ruts and if you listen closely enough hear the steps. See Grand Lakes News Online website and Fugate (1991) for more Military-Texas Road information.

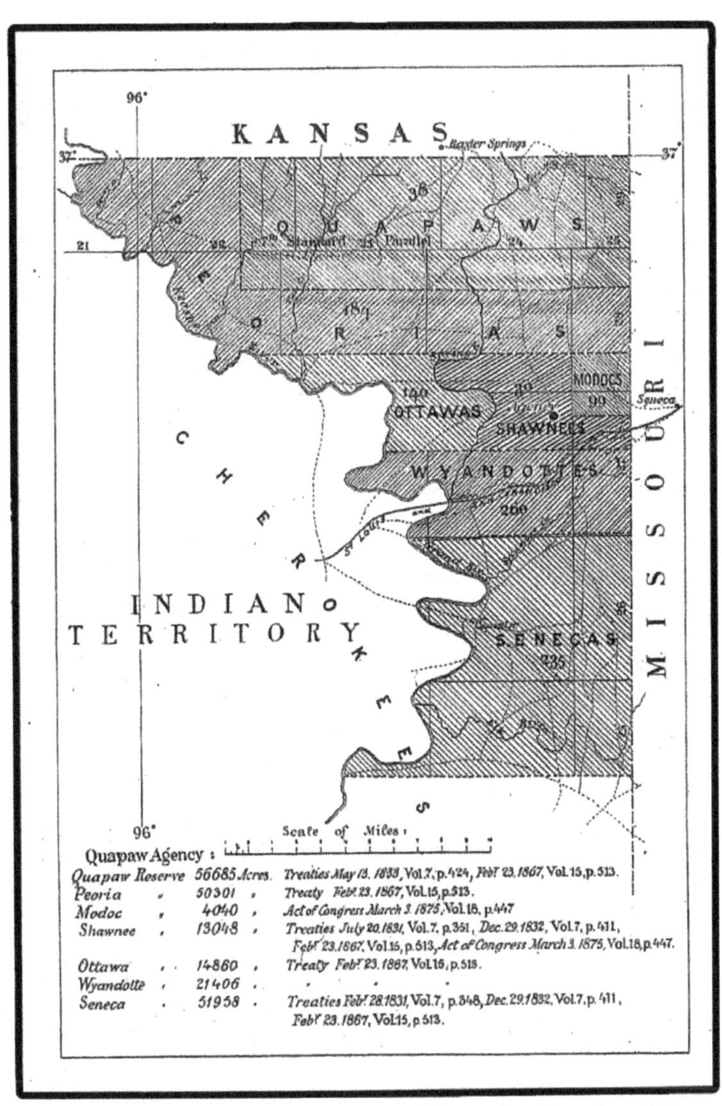

An 1879 Quapaw Indian Agency map with treaties. Found on E Bay.

Halahtookit—Daytime Smoke—Son of Clark—Me Clark. Nez Perce son of the famous explorer William Clark. Halahtookit and his tribe were exiled to Indian Territory-Quapaw Agency as prisoners of war. Photo courtesy of the Minnesota Historical Society.

The Story of Halahtookit
Nez Perce Son of William Clark

1879
Quapaw Indian Agency, Indian Territory

some say there's ghosts
in these parts
might find him wedged up
in tree branches
his reddish hair
and eyes like the sky
confuse folks around here
and he won't be forgotten

came here in '78
called him **Halahtookit**
Son of Clark
a Nimiipuu prisoner of war
exiled to Indian Territory
Quapaw Agency
after the chase almost to Canada
his tribe loses at Bear Paw
government councils punish them
send them to IT

first to Fort Leavenworth
Eekish Pah
The Hot Place
malaria already killing 'em
then to Baxter Springs
five children gone and buried
by the depot
hungry hogs yield up the graves
before they even leave

sent off to Quapaw Agency
agent counts them 411—*eighty-three adult males*
one hundred eighty-four adult females
one hundred forty-four children
squats them on Modoc land
where they figure they can
break, tame, or civilize wild urchins
lead them *to civilization*
by '79
down fifteen adult males
fifty-seven adult females
twelve children

now he's gone
but listen careful

this story I am telling you
won't be forgotten
it will stick in your craw
set you thinking
right and wrong
how nobody wins
and we are all prisoners of war
when we do to the least of us

In 1879, the son of one of America's most famous explorers, William Clark, was exiled as a Nez Perce (Nimiipuu) prisoner of war to Indian Territory-Quapaw Agency with Chief Joseph and the rest of their tribe. They were squatted on Modoc land where the government hoped an already defeated tribe could help civilize Indian Territory's newest exiles.

See Osborne (2017) and Pearson (2008) for further understanding of Me Clark and the Nez Perce experience in Indian Territory.

John Patrick McNaughton: A Civic Minded Kind of Man

1932
Miami, Oklahoma

these days

you'd probably call me

high rolling

entrepreneur

civic minded kind of man

married Chief Batiste's granddaughter

got proper ruling to incorporate

first town in Indian Territory

worked my way to digging rights

on restricted land

by 1892, hauled out 3 million tons

of lead

12 car loads of zinc to Joplin

look at my place

three and one-half story barn celebrated at the World's Fair

800 acres in crops and tamed grass

first system of private water works in the IT

brought first spool of barbed wire

to Peoria Indian land

I could go on and on

don't want to sound too much a bragger

but here is the deal

some folks just pass through

I acted

you go after what's right there

or you and everything else

just dies obscure

I chose progress, I chose prosperity

John Patrick McNaughton was one of Ottawa county's (Oklahoma) most well known citizens—what the *Indian-Pioneer History of Ottawa County, Oklahoma 1937-1938* interviews stated, " a most useful of the county's pioneers." See Cook (2012) for further information about McNaughton. After many refusals, he was the first white man to get permission to start prospecting and then mining lead on Peoria Indian land. Mr. McNaughton was not a man to cross—his words are still visible in his celebrated barn. "Thieves and prowlers stay out of here. I will pay $25. reward for his undertaker." (personal communication—Sandy Ankenman)

A 1914 photo of the Davis Mining Company in Peoria, Oklahoma where some of the first commercial lead mining took place in Oklahoma. Bumper Ferris, runaway Seneca Indian School boy, is standing with his hands on his hips, in a white shirt, and a seemingly new hat. Photo provided by Arnold Richardson.

Runaway Seneca Indian School Boy Bumper Ferris

1914

Davis Mining Company Peoria, Oklahoma

see him there

hands on his hips

feelin' pretty good

new hat and all

but here is the thing

he may look small

but he's working

a real man now

puts in a fair day

learnin' this mining business

no more running from that crazy school

where they took his clothes

and judged they could stop him

just made him faster

watch him slide past Lost Creek

head for Peoria

now he works like the man

he knows he really is

no more school

no more can't talk his language

no more take him 'way from kin

they think they can keep him down

he knows way better

Ottawa County had several Indian schools including the Seneca Indian School founded as the Wyandotte Mission by Quakers back in 1868. The Bureau of Indian Affairs had established how little Indian children should be educated.

> To this end only English should be allowed to be spoken and only English-speaking teachers should be employed in schools supported wholly or in part by the government. In government schools, no textbooks and no oral instruction in the vernacular will be allowed, but all textbooks and instruction must be in the English language. No departure from this rule will be allowed except when absolutely necessary in rudimentary instruction in English. But it is permitted to read from the Bible in the vernacular at the daily opening of school when English is not understood.

See Niederding (1983) for information regarding education in Ottawa County, Oklahoma.

An old man gave me the picture of the Peoria mine workers and told me the stories of the men in the picture. His father in law worked in one of the original mines and was later killed in a mining accident. Bumper Ferris was the young Indian boy who kept running away from the Seneca school where they kept taking his clothes. (personal communication—Arnold Richardson)

A Quapaw By Any Other Name
Mid-1600s
Mississippi River

appears some believe

a name is just not that important

but I been thinking real hard

and I know better

the way memory fixes on you

won't let you go

so listen up

these folks are **Quapaw**

Downstream People—O-Gah-Pah

Ugaxpa—Ogaxpa—Okaxpa

but way back they were part of **Dhegiha**

and one day the whole bunch

goes traveling across the big river

when an ol' grapevine broke

and their people went downstream

and after all this time

end up in Indian Territory

but here's what you need to know

Quapaw had it figured

maybe even realized

all of us are downstream

whatever happens upstream

going to get all of us
and the sooner we understand this
the better we take care
what we call home

There are many stories of how Quapaw—O-Gah-Pah—Ogaxpa—Ugahxpa—Okaxpa— Arkansea—Downstream People got their name. It's thought at one time Quapaws, Osages, Poncas, Kansas, and Omahas were one tribe. One story tells how they were all migrating together when they got to a river, the Quapaws were slow and when they crossed, they didn't know which way the others had gone. So the Quapaws went downstream. Another version is when they were crossing the Mississippi, they used a vine rope to cross and the vine broke. By the time the rope was fixed, it was foggy and they were lost...so the Quapaws went downstream while the rest of the tribes went upstream. See Oklahoma Indian Affairs Commission (1977) first person Quapaw stories.

In today's world, environmentalists remind us, "we are all downstream." Wendell Berry tells us to "do unto others downstream as you would have those upstream do unto you."

Six Harmless Idiots at the Quapaw Agency
1897
Quapaw Agency-Indian Territory

and they've all got the heebie-jeebies
'bout what's in the makes for them
peculiar how they know
and how it doesn't sound good
you wonder how they got it figured
well, you start to understand now
if you think hard enough
someone is bound
to hear you
and in some kind of mysterious way
what happens
to any of us
finally finds home
again in the rest of us
so those boys just linger
and listen to what sounds like
digging and rumors
swirling lead and zinc
crazy dreams
right smack there at Quapaw Agency, Indian Territory
but listen
we white folks got ourselves

this here crystal ball

it fortune tells

predicts yours, my destiny

you think we'll see orange water

blowing chat

sink holes ready to swallow us

even the ol' U.S. of A government declaring disaster

tear down, remove whole families

even a town

and a tornado whipping through

ever hear that ol' expression

when the chickens come home to roost

'cause we're all scared

and it's not going

to be pretty

Prior to opening a separate asylum for "insane" Indians, a report was requested "as to the practicability and utility of such an institution." Commissioner of Indian Affairs W.A. Jones sounded out Indian agents and reservation superintendents. The Quapaw Agency in Indian Territory was reported to have "six harmless idiots." See Joinson ((2016) for extensive information re: the Canton (Hiawatha) Asylum for Insane Indians.

Father - Nau-he-gon-sa
Mother - Mah-tah-we-h-see

Tribe: Shawnee
File 973

QUAPAW INDIAN COMPETENCY COMMISSION

C. L. ELLIS, SPECIAL INDIAN AGENT
J. F. MURPHY, QUAPAW COMPETENCY COMMISSIONER
IRA C. DEAVER, SUPERINTENDENT QUAPAW AGENCY

Raised to 4/4

#973

EXAMINATION OF ALLOTTEE

Allotment No. 58 — Degree and kind of Indian Blood 1/4 French 3/4 E. Shawnee — 1910 Census No. 11

What is your name? Captain, Thomas, Sr.
Postoffice? Seneca, Mo — County of Ottawa (Resides in) — State Okla
When born and where? About 1850, Quapaw Res — Married or single? Married
1st wife — Skaw-wan-ah-dee Davis (Seneca) — 2nd wife — Tee-dee-day (Shawnee) (Bluebird)
3rd wife — Martha Ellen Gullat (white) — Nationality

Allotment No. — CHILDREN — AGE, Date of Birth

Died Jan 10, 1904 (by 1st wife) Henry Captain — Born 1870
Minnie Captain (Dead) no record
Thomas Andrew Captain — Dec 21, 187?
Cordelia Jane Captain (Hampton) — July 9, 18??
Mary Ellen Captain (Woodall) — Feb 8, 18??
Died Dec 13, 1905 Charley Selby Captain — Aug 14, 18??
Sarah Mary Captain (Crane) — May ?, 18??
Frances Captain — Nov 24, 1885
Grace Elizabeth Captain — May 16, 1892
George Alexander Captain — June 17, 1899
Martha Eveline Captain — May 26, 1901
Sophrona Anna Captain — July 2, 1905
All children allotted except the young-est child

Do you reside on your allotment? No — If not, where? Martha Eveline's allotment
Do you desire restrictions removed from all but 40 acres (homestead)? Yes
Why do you wish the restrictions removed? Want to sell
Why do you not wish the restrictions removed?
If the restrictions are removed what will you do with your land? Sell it, get money pay debts at store and bank
Have you any inherited lands? Yes
How much is your interest in these inherited lands? 1/2 in Charley's allotment 160 A
Have you ever disposed of your interests in inherited or trust lands? No
If so, what did you receive for the inherited lands? $
For what did you use the money? Improvements on Martha Eveline's land, consists of house, stable, sheds, worth about 100.00
To whom and for what amount did you sell your allotted lands? None sold
For what did you use this money?
Mr. & Captain and the boys farm 20 A on the father's land
How much land in your allotment? 160 A
Do you farm it yourself or lease it? Lease it — How much do you cultivate?
To whom is it leased for agricultural purposes, how long, and how much do you receive? B. Sparlin, 53 A at $100.00 per yr, & more pro to run
To whom is it leased for mineral purposes, how long, and how much do you receive?
Are rentals collected in advance? Yes — If so, how far? one year
Do you rent other lands? No — If so, whose and how much do you receive?
Charley's allotment is rough, rocky land in timber
Who assisted you in making your leases? B. Sparlin
Have you made permanent improvements on your land? Yes — If so, to what extent? Barn and fence
Cost of improvements? $ 400.00 — What source derived? Rentals
How much is your land worth per acre? 40 A worth $50.00 per A, and balance do not know
Why do you think it is worth that price? 40 A good prairie land

Thomas Captain's Quapaw Indian Competency Commission examination of allottee's papers (pg 1).

What is the most you have ever been offered for this land? $ 50.00 for prairie 40 A
Who made the offer and when?
Have you signed a contract or made an agreement of any kind to sell this land when restrictions are removed? No
With whom and at what price?
Have you borrowed any money expecting to pay it from sale of this or other land? No
From whom?
Have you a personal bank account? No What is your present balance approximately? $
Do you pay cash for your supplies when you get them?
~~If not, how often do you pay?~~
Are you in debt? Yes How much? 600.00
To whom? Norris & Norris and Bank of Seneca For what? Groceries and clothing
Have you been in debt before? Yes How did you settle? Rental
Have you borrowed any money? Yes $200.00 From whom, amount, and on what security? Chattle mortgage 2 horses and 2 mules, due next month.
What business experience have you had? None
Give name of two business men who know you well? Norris & Norris, Seneca, Bank of Seneca "
To whom do you go for advice in business matters? B. Sfarlin
What property have you besides your allotment? Two horses and Two mules.
Do you use intoxicating liquors? Yes To what extent? Some times as often as he can get
Did you ever use them, and if so, when did you quit?
Do you use the mescal bean? No To what extent?
Are you in good physical condition? Yes Do you work? Yes
At what? Farm little How much do you receive per $
Have you attended school? No Where and how long?
Do you send your children to school? Yes Where? District and Seneca Bo[illegible]
Why not?
Do your children do any work? Yes How much? Like Indian
For whom? What pay do they receive per
How many are there dependent on you? Wife and six children at home
Do you support your family in as good shape as your neighbors? As well as Indian neighbo[rs]
Occupation of husband and salary?
Old man is in bad shape financi[ally] In debt and rents collected ahead & M[ortg]age coming due on stock and no pro[perty] to dispose of to meet the obligations & Drinks [illegible]

OTHER FACTS AND NOTES:
LEASES—If Mining, Extent of Development and Production, Term, Value, Etc.

Lessee	Postoffice	Kind	Period	Acreage	Rental

allotted 160 acres
Restrictions Removed from 40 acres Conditionaly Not sold to [illegible]

Recommended for 2nd Class. 11/18/10
(Date ending) [illegible]

Thomas Captain's Quapaw Indian Competency Commission examination of allottee's papers (pg 2).
Note some of the questions—"Do you use intoxicating liquors?" "Do you use the mescal bean?" "Do you send your children to school?"
Papers provided by Jackie Captain Brewer

Incompetent Restricted Quapaw Indian about 50 Years Old

1924 Griffin Cemetery
Quapaw Agency, Oklahoma

now you see him

now you don't

there he is

lined up with the old Quapaw men

still had his hair those days

all smoky black braids

look careful again

see his fist

hold that old lump

of lead

he carried it back then

made him rich

but now they call him crazy

incompetent

give him a guardian

send him to Hiawatha

Insane Asylum for Indians

way up in South Dakota

no more money
no more self

do-gooders write a book
Oklahoma's Poor Rich Indians
An Orgy of Graft and Exploitation
get folks all stirred
close down Hiawatha
town fathers buy up the place
build themselves a golf course
right over his crazy dead Indian
brothers and sisters

back to the Agency
mute now
lips sewed tight

now you hear him
now you don't

"Robert Thompson, an incompetent restricted Quapaw Indian about 50 years old. He was sent to an Insane Asylum when the present guardian took charge of his estate, a little over two years ago, $24,000 was receipted for. The Liberty Bonds and all securities have been disposed of, and the balance on hand (November 1913) amounted to $54.40."

Thompson was sent to the Hiawatha Asylum for Insane Indians in Canton, South Dakota. Hiawatha was one of two federal asylums in the United States. It was closed in 1934 following investigations into horrendous conditions. Hiawatha was designed specifically for America's Indian population. When the asylum was closed, the town of Canton built a golf course over the graves of patients who had died there. See Zitkala, Fabens, & Sniffen (1924) , Joinson (2016), Leahy (2009) for a better understanding of "Canton-Hiawatha Asylum for Insane Indians." Also, Nerburn (2013) for an account of a young girl sent to Hiawatha and her family's search for her story.

REGULATIONS TO BE OBSERVED

63981

IN THE

LEASING FOR MINING PURPOSES OF ALLOTTED LANDS OF INCOMPETENT INDIANS OF THE QUAPAW AGENCY, INDIAN TERRITORY.

PRESCRIBED BY THE SECRETARY OF THE INTERIOR, JANUARY 24, 1907, FOR THE PURPOSE OF CARRYING INTO EFFECT THE PROVISIONS OF THE ACT OF CONGRESS APPROVED JUNE 7, 1897 (30 STAT. L., 72).

WASHINGTON:
GOVERNMENT PRINTING OFFICE.
1907.

Cover of a 1907 Government Printing Office document. Retrieved from the Library of Congress.

"New Rich" is a 1939 lithograph by Charles Banks Wilson (1918-2013). The Quapaw couple is shown here with their new car obtained from lead and zinc royalties. Permission given by the Gilcrease Museum.

Real Type of Old Full-Blood Indian
Hardly of Average Intelligence

October 12, 1920
Quapaw Agency, Oklahoma

she was one who survived

government says

real type of the old full-blood Indian

and is hardly of average intelligence

and is very poorly informed

needs protection

so she stays low

watches others live their money rich

fancy car lives

so scoffed at by us white folks

but she keeps her ways

cooks platter corn, bean bread, grape dumplings

grows a big garden

believes her knowledge

will be figured right and good

maybe even more valuable

someday

than all that zinc and lead

By 1926, some Quapaws were among the wealthiest people in the world. An 1897 federal statute had empowered the Downstream People to lease allotments for mineral purposes without government supervision for a period of ten years. However, the Quapaws seldom obtained the full value of their inheritance. Often lawyers or trusted associates had themselves appointed guardians of Quapaws declared "legally incompetent." Eventually, government intervention led to hearings and federal authority over leasing.

On January 8, 1921, the Commissioner of Indian Affairs was given a survey of the Quapaw Indians by the Department of Interior and the United States Indian Service. Descriptions of the allottees ("She never attended school, does not speak English and is a real type of the old full-blood Indian. She has no business ability whatever and she is not qualified to handle her property and business affairs successfully herself.") their Quapaw and financial status, and the actual land allotments were included in the survey-report. See United States Congress House Committee on Indian Affairs (1919) and the Government Printing Office (1907) for an understanding of the relationship between the federal government and the Quapaw tribe.

Peyote—Devil's Root or Plant of God?
1890s
Quapaw Agency, Oklahoma

they think I'm magic
this earthy essence of mine
I assure possibilities
when everything seems bleak
I will dance in your brain
make dreams touchable
scare evil
away

John Wilson or "Moonhead" brought peyote worship to the Quapaws in the 1890s. Its message of hope and belief that if peyote was eaten reverently and with good thoughts, their sick could be brought back to health. Native American Church services consisted of songs accompanied by drum and rattles, prayers, the use of peyote as a sacrament, and the sharing of fruits and candies. The new religion was an explosive topic among missionaries and Indian agents and many called peyote the "devil's root." See Nieberding (1976), Baird (1980), and Oklahoma Indian Affairs Commission (1977) for an understanding of the Native American church in the Quapaw community and white reaction.

A Ku Klux Klan gathering in Douthat, once a booming mining town, now a ghost town. Note the signs—"Protect the American Home Our Children Our Home" "American Order." Photo courtesy of Orval "Hoppy" Ray estate.

Look How My KKK Brothers Love My Light
May 7, 1927
Quapaw Indian Agency, Oklahoma

face it
you know you, too, can't stay away
me, a cross, up on the chat pile
burning all bright
the Indian school lit in my shadow
my white-robed brothers
rallied and ready
and you, too, know
how righteous those St Mary of the Quapaw
black- hooded sisters
and their little catholic wannabe's
want to sound
and how I know and you know
they just better learn some respect

yeah, my embers will turn to dust
but I promise you
right here and now
I will blow like chat
invade your pores, their pores
and me and my fire
won't be forgot

The May 8, 1927 *Miami Daily News-Record* reported a Ku Klux Klan cross burning on a chat pile near the St Mary's of the Quapaw (Indian) School. Years ago, I met an old man walking down Main Street in Miami. We got talking about Miami's history and I asked him if he had any memories of the Ku Klux Klan in these parts. He told me when he was young he had heard about a cross burning out near the Indian school—said he went out there the next day and just found the embers.

In 1915, drilling operations led to the opening of rich mines on the site of Picher and was originally Quawpaw Indian land.
Photo courtesy of Ed Keheley

Tri-State miners with their "sunshine lamps." These pre-OSHA oiled lit wick lamps provided illumination in the dark mines. Photo courtesy of Ed Keheley.

Tri-State miners—"Hookers" sending loaded cans up. The shovelers were responsible for filling the cans with 1,250 pounds of ore. In an eight hour shift, a shoveler normally loaded twenty-two tons. Photo provided by Bill Crawford.

The Shoveler
February 2, 1939
Douthat, Oklahoma

someday you'll find me
all written up
in the *Miami Daily News-Record*
maybe even all permanent bound in
the *Killed and Injured in the Mines*
Ottawa County book
but today just see me
down deep
doing what I've been doing
what seems like
some kind of forever
yeah, I got pride
you don't get these kind
of muscles
just sitting on your butt
I fill those cans
1,250 pounds lead-zinc full
more I send up, the more my pockets
get filled
my wife she's all worried
thinks this old cough of mine
is getting worse

but I got mouths to feed

and I will keep going down

when your time is up

it's up

don't make no difference

fretting about it

today is just another day

who knows any way

about tomorrow

"Toiling Miners Continue Search for Two Bodies After Recovering Three, Weary Shovelers Valiantly Carry on Quest at Dines Southern Mines, Two Injured Survivors Recovering at Picher, Okla" was the headline in the *Miami News Record* on February 2, 1939. Mrs.
Margaret Parker spent four years researching the two volumes, almost 500 pages record, *Killed and Injured in the Mines of Ottawa County*. (2006) She grew up in Picher and her father started working in the mines when he was 13 years old.

Miner's Daughter Comes of Age
1919
Cardin, Oklahoma

my mama
be gone seven years
found her all bloodied
a crochet hook up
her privates
nosey church ladies descend
cut up her ol' dresses
making me all new garb
like they doing me a favor
don’t want that kind of remember
still can't stop the picture
just keeps rewinding
blood everywhere
no more brothers, no more sisters

now I'm twelve
got myself all this responsibility
and I hear the whistles blow
three short, twelve in a row
everyone knows what that means
and me
can't figure how I certained

just knew

when they carried Daddy

up the shaft in that awful cage

looked all around me

saw them wide-eyed wives

whispering rumor chatter

and then the hoistman brung up

his crew

they won't even look at me

just heard them mumbling

one tiny rock

saw his chest rise and fall

rise and fall

and I get to praying to Jesus

I sit vigil

four forever days

when his chest goes flat

you think they was scared leavin'

worried 'bout their papa-less

mama-less child?

An old family friend told me her mother's and grandmother's story. The health conditions in the mines and mining towns made bringing another child into the world daunting. (personal communication Oma Potts)

Pebble Turns Killer in Cardin Mine
1919
Cardin, Oklahoma

you think I'm tiny
but here I've got the upper hand
so to speak
and when I ping your brain pan
you might think
before you don't no more
how something
so small
can't do no harm, make no change
but let this be a lesson
intention, aim, velocity
make all the difference
and really any of us
have this power deep inside
me, I just used it
and you paid the price

When the old woman first told me the story of her mother, grandmother, and grandfather, I understood that a huge slab had crushed her grandfather. A few years later, the great granddaughter told me that actually it was a tiny rock that hit her great grandpa. As I listened to more mining stories, I learned how tiny rocks were feared as much as big slabs or cave ins. (personal communication Retha Epperson)

William Hezekiah Cantwell and Nellie Wiles Cantwell playing at a barn dance circa 1930s. He farmed and worked in the mines, she worked at the Quapaw Indian Agency and their farm.
Photo provided by Nellie Cantwell.

William Hezekiah Cantwell at the Oregon State TB Sanitarium

1949
Salem, Oregon

now I lay
me in scrubbed white sheets
the south window
brings as much sun
as an Oregon sun can do
my eyes never quite right
to light
after all those years
underground
I remember the kids
how they swung
from my stretched-out arms
how I could pick and shovel
all day
and still fiddle
at the barn dances come night
how finally it was time
to get out
before the dust
ate me up
we packed the old rattle trap

with mattress on top car
Okies headed for Oregon
where green became my Nellie's
favorite color
now I cough red
they think my oldest daughter
got this TB, too
and her with a new young one
that dust
it just keeps chasing

If you were fortunate enough to escape falling slabs, tiny rocks, and overturned ore buckets you were likely to develop a case of what the miners called 'miners con'—an almost death sentence. Silicosis led to decreased working ability, coughing, weight loss, weakness, and respiratory infections, including tuberculosis. In the 1930s, Ottawa County had the highest tuberculosis mortality rate in the United States. See Rosner and Markowitz (2006) and Gibson (1972) for information regarding health conditions and mining in the Tri-State Mining District.

This 1932 photo shows the rescue attempts for three year old Gerald Collins who had fallen into a mine shaft near Peoria, Oklahoma. Possibly the inspiration for L.S. Davidson's depiction of "Dim-Wit" in the 1939 muck-raker novel, South of Joplin: A Story of Tri-State Diggin's. Photo found on E Bay.

PICHER, OKLA. - OKLAHOMA NATIONA GUARDSMEN WERE CALLED TO PICHER, FOLLOWING SERIOUS DISORDERS BETWEEN UNION AND NON-UNION MINERS, IN THE TRI-STATE ZINC MINE STRIKE. STATE MILITIAMEN ARE SHOWN OUTSIDE THE CONNELL HOTEL, OWNED BY F. W. EVANS PROMINENT MINE OPERATOR AND HEAD OF THE NON-UNION ORGANIZATION. SENT A AND C LISTS CLEVELAND TO

National Guard troops called out to Picher during the 1935 strike. The bottom photo shows Blue Card members, a company sponsored union. Note the pick handle axes—a symbol of their union. Photo courtesy of Orval "Hoppy" Ray estate.

Mule Water Boy Hears Strike Talk

May 21, 1935
Picher, Oklahoma

been called *Dimwit*
for how long
story goes
I fell head first
down that ol' mine shaft
folks say
never was the same
like I lost my ears, too
but I got ears and ears
that listen hard
and the mules they trust me
I keep 'em in water
you see, the mules and me
we know stuff

I hear them Blue Card boys talk
smell blood
know they think pick handles
will clean up this strike business
company boys really
think they can scare folks talkin'
about foreigners trying to tell you what to do

you think about it
big ol' Eagle-Picher making some big money
while down deep
folks keep chancing life and limb
shovelers making 11 and 1/2 cent a 1,200 pound can
powder men loading dynamite
sure all the fuses lit
then last man out the ground
look up you see the roof trimmers
hanging up there one hand on the ladder
the other prying off loose rock

yeah, the mules and me
we know secrets
when I come out to light
I'm smart on all that stuff
you thought was hid
and someday
you might just wish
you'd shown some respect

At midnight on May 9, 1935, the International Union of Mine Mill and Smelter Workers called a strike seeking "better working conditions, a shorter work week, and adherence to American standards of a living wage." Almost all of the 55 mines and tailing mills then operating were shut down. More than four thousand miners and workers were involved. The Blue Card Union, a back to work company-sponsored union, was created in reaction to the strike. A pick handle was their symbol and pick handle parades were used to intimidate strikers and their sympathizers who they believed were led by "radicals and reds." Eventually, violence broke out—including a mob assaulting union members and their headquarters and the attack of three law enforcement officers. The Oklahoma National Guard was called in. During their 1934 summer encampment, emphasis was placed on teaching techniques for containing "mobs, strikes, hunger marches and demonstrations." Suggs (1986) is probably the most complete account of the 1935 strike.

Dimwit was a character in the muckraker novel, *South of Joplin: A Story of Tri-State Diggin's* written in 1939 by L.S. Davidson. "The story told about his life was a familiar one in the mining district. As a child he had been thrown in a shaft, and one year after the accident both his parents died with 'miners con.' (He) was thought to be an idiot by the miners."

Ottawa County's First Public Health Nurse
Starts to Wonder

1926-1949
Picher Mining Field

for twenty-three years

I did the work

first public health nurse

in the mining district

Tri-State Zinc and Lead Ore Producers Association

my employer

back in '26 typhoid, smallpox, diphtheria

raged

later drove how many kids

down to Talihina TB Sanitarium

kept pounding education

an ounce of prevention a pound of cure

could hardly convince those hard headed miners

Depression come along

Ore Producers furnished shoes for 503 kids

the men went out on rabbit hunts

clubbin' them out in the fields

day old bread and rabbit

at least

they ate and had shoes on their feet

some say I **was** the welfare department

don't know about that

you're a nurse

you do what you gotta do

but I think about all that bad breathing

report came out in '52

the greatest hazard lies in the failure of individual miners

to use proper methods

I'm starting to wonder

guess I'll never really know

Living conditions in the mining district were pitiful. In 1939, *A Preliminary Report on Living, Working, and Health Conditions in the Tri-State Mining Area* was published. Dilapidated housing, chemical-filled mill ponds, expanding chat piles, and economic hardship were what you just coped with. One miner told the investigators, "It's a pretty unhealthy place to live, but the chemicals in the mill ponds fertilize the water and make the flowers grow." Folks said when the Great Depression started people knew there was a depression because there was just a little more hard times than usual. Ruth Hulsman was the district health nurse for the mines and was employed by the Tri-State Zinc and Lead Ore Producers Association. See Dick (1940) for a documentary about Tri-State living conditions, and Rosner & Markowitiz (2006), and Gibson (1972) books for an extensive understanding. See Nieberding (1976) for Hulsman background.

Mortician's Wife Fixes Them Up Nice
1949
Picher Mining Field, Oklahoma

when the whistle blasts

three short

twelve in a row

you know accident, fear

disaster

cave-ins, slabs falling, dynamite explosions

somebody trapped way deep

I watch my man

and always another

afraid one might pass out

when they see the mangled

then get straps ready, the wire basket

go down now

then pull 'em up vertical

dead or alive

screech their way to Picher Hospital

or back to me

where we fix them up nice

you know, honey, you got to be an artist

and you gotta care

By the 1930s, Picher had four funeral home-ambulance services. Quite the busy and lucrative business with all the mining accidents and fatalities. (personal communication Wanda Thomas)

Picher in 1950. Photo courtesy of Ed Keheley.

Picher in 1950 when 200 homes and businesses, four blocks off Main Street, were closed off due to an impending cave in. The photo shows a sign warning "surface may cave anytime—enter at your own risk—area extensively undermined." Town fathers had proposed a bird sanctuary for the area. Photo courtesy of Orval "Hoppy" Ray estate.

Picher Screech Owl Gives Three Screech Warning

1950
Picher, Oklahoma

back in 1950

town fathers thought themselves

pretty smart

four blocks off Main Street

200 homes and businesses get done away with

lookin' like a subsidence

ready to happen

you know cave-in, sink hole

fence off the place

then decide a bird sanctuary

would be real nice

me, I do my three times screech

almost like the mine whistles

tellin' bad

Quapaw folks, they smart

hear my screech

know impending death

when they hear it

you'd think

someone woulda heard

when something could still be done

Quapaw people know to listen to owls because other beings talk through them. A screech owl calling three times warns of impending death. In 1950, a four-block area of downtown Picher showed signs of collapse. Town fathers fenced off the area and a bird sanctuary was established. See Fugate (1991) and Oklahoma Indian Affairs Commission (1977) for more information.

Picher at two years old in 1917...now a booming mining town. Photo courtesy of Ed Keheley.

Another 1917 photo of booming Picher during World War I when most of the lead for the American war effort was being produced in the Tri-State Mining District. Photo courtesy of Ed Keheley.

Picher in the 1930s. Photo courtesy of Ed Keheley.

Bonnie (Minus Clyde): A Poet and Prophet
May 23, 1934
Bienville Parish, Louisiana

if you reading this

it's all hind sight

and I can't even figure how I'm getting

this all communicated to you

but back a few months we shoot dead

two officers out in Grapevine

and end up off this here Texas Road

seeing all these mountains of chat

lead money to be found

and get ourselves stuck in mud

try flagging down help

but all we get is Police Chief Boyd

and Constable Cal Campbell

and a big shoot out

right there in Commerce, Oklahoma

take out Campbell

kidnap the chief

me, time ticking away

month later

pay back, they shoot me up dead

you think I'd do any different?

probably not

but listen up a little more
bet you didn't know I was prophet
and poet
sent mama this poem

someday they'll go down together
and they'll bury them side by side
to a few it will be grief
to the law a relief
but it's death to Bonnie and Clyde

On April 16, 1934, Bonnie and Clyde kidnapped the Commerce, Oklahoma police chief and killed a constable. Some thought they were notorious criminals, others believed they were modern day Robin Hoods. Bonnie's poetry may have added to her supposed charm. Weeks before being gunned down on May 23, 1934, she sent her mother a poem— "The Story of Bonnie and Clyde." It predicted their demise. See the web site The Texashideout.tripod.com for Bonnie Parker's poems.

Country Boy Meets the Cookson Hill's Robin Hood

February 18, 1932
Wyandotte, Oklahoma

feeling pretty special now

but yesterday

no so much

actually darn spooked

fancy car pulled up right there

in front of me

looked like they was lookin'

for directions

the best way away

got me thinking

this wasn't just nobody

Picher paper got it confirmed yesterday

now townies all a titter

yeah, Pretty Boy Floyd

right here in Ottawa County

and me sent him safe

you know that Woody Guthrie

even got a song about him

something 'bout some rob with a six-gun

others with a fountain pen

well, folks, I know who my hero is

and I got to say

right here and now
I got my wits about me
no more fraidy-cat
'cause it's time and I am ready
so open your peepers
and watch them mortgages burn

My mother, grandmother, aunts, and uncles could sit up all night and tell stories. I never got tired of hearing my uncle Dick talk about his encounter with Pretty Boy Floyd. (personal communication Richard Cantwell) The February 18, 1932 *Picher King Jack* newspaper reported him coming through Picher that week. During the Depression, a lot of folks called him the "Robin Hood of the Cookson Hills." When he robbed banks, he destroyed mortgage documents which freed many citizens of their debts. See Cray (2004) for background on how Pretty Boy Floyd became both legend and song.

A Rumor of Panthers
May 28, 2000
Miami, Oklahoma

don't ask me how I know this

but I hear the boys

down at the feed store

say I'm out there

say they've heard me

yeah, a panther, can't mistake that screech

I will stay low

keep my mystery

my paw prints lost in dust and water

Panthers have long been significant to Quapaw people. The Minneapolis Institute of Art has a Quapaw vessel circa 1500 signifying an Underwater Panther. It's been said a man on a vision quest might well be happy if he dreamed a panther. Years ago, I was in the Miami Feed Store ("if it swims, walks, or crawls, WE FEED them all") and eavesdropped on some farmers talking about hearing a panther. One of the last native Quapaw speakers told me about seeing huge cat prints along the Spring River. (personal communication-Ardina Moore)

An Eagle-Picher advertisement from World War II. Statistics vary but most of the lead for American military use during the war came from the Tri-State Mining District. Advertisement found on E Bay.

Brother Lead and Cousin Zinc Want to Know

2016
Tri-State Mining District-Oklahoma, Kansas, Missouri

both of us now
start to question
just what if
we stayed underground, laid low
how what if the ol' U S of A government
hadn't sent down whole bunches
of brown-skinned folks
to a land nobody wanted
and what if
our mineral perfect selves
hadn't been just right
for ammunition, medicine, batteries, paint
makeup, print, gas all the stuff
people now just feel a right to
but there we were
ripe to be discovered
and suddenly we're a *Wilderness Bonanza*
everyone wants a piece of us
wants back what first nobody wanted
all the finagling, swindling
but we're used up
look at us

orange water, sink holes, chat piles
we left our legacy
but, oh, Lord
what will become of any of us
now?

Lead in northeastern Indian Territory played a part in the Civil War when both the North and the South fought for control of IT and the minerals they could use for ammunition. By the early twentieth century, Picher, Oklahoma was the largest mining town in the Tri-State Mining District (northeastern Oklahoma, southeastern Kansas, southwestern Missouri). During World War I, it's estimated at least half of every bullet fired by an American was made of lead that originated in Picher. Similar statistics held true for WW II. Gibson (1972) and Robertson (2010) provide well-researched history and geology of the Tri-State Mining District.

Blowing in the Wind Chat Pile
August 31, 2009
Picher, Oklahoma

they list off my elements

cadmium

lead

zinc

I'm the left over

blowing in the wind

answer

my friend

I slip through your pores

take hold of you

maybe even grow

to become one

they say I am toxic

but we've been neighbors

for so long

you watch me out your window

make sand boxes for the kids

leap down, sled down, 4 wheel down

my crazy declines

fill your driveways and roadways

you think will carry you away

but we're home to each other shared too much too long

to know
how to leave now

You can still see the chat piles scattered in and around Picher, Oklahoma. Chat piles are giant piles of toxic metal tailings, some as tall as 200-300 feet. Chat is the fine gravel waste left over after ore production in zinc and lead mining. They contain high amounts of lead, zinc, and cadmium that can leach into groundwater, be eaten by livestock, contaminate local wells, and be blown from piles to neighboring areas. The adverse health effects are directly related to the contaminants that exist in the chat. For an understanding of chat-mine tailings, see Gibson (1972) and Jim and Scott (2007).

Tri-State miners—"Screen Apes" who were responsible for breaking large chunks of rock to smaller pieces to fit through the screen. Mickey Mantle reportedly worked as a screen ape where he developed his powerful (baseball bat) swing. Photo courtesy of Orval "Hoppy" Ray estate.

Big Fish Mickey Mantle

August 30, 1995
Dallas, Texas

pretty amazing, huh?
Big fish in a small pond
becomes big fish in the big ol' ocean
but first a rookie
almost stuck in a slump
when the old man
tells me
I thought I raised a man.
I see I raised a coward.
You can come back to Oklahoma
and work the mines with me
but now I'm the *Commerce Comet*
and I left you all behind
no going down deep
eating dust
the way my pa dead already
at 39
but some miseries just keep following
lust, booze, sour money
but still I had my glory days
MVP x three
twelve, count 'em, twelve world series

seven championships

not much time left now

but I got saved

and I remember when that ball

went chasing over the river up at Baxter Springs

knew my ticket spelled, *go*

and I went

Mickey Mantle, the Commerce Comet, is probably the Tri-State Mining District's most famous hometown boy. Both he and his father worked in the mines. Story goes that Mantle developed tremendous strength in his wrists, shoulders, arms, and forearms working as a "screen ape." Screen apes were responsible for smashing large rocks into small stones with a sledgehammer.

Leavy (2010) chronicles Mantle's life including growing up in the Tri-State Mining District and the discovery of high lead levels in his blood at the end of his life.

Picher in 1983. Note the chat piles. Photo courtesy of Ed Keheley.

Chat Rats Reminisce the Good Ol' Days
2000
Picher, Oklahoma

yeah, those were good times

every day

all day

up and down the chat piles

then down to Tar Creek

swim past floating dead dogs don't scare us chat rats none

then up to that old junked fridge

right smack in the middle of the creek

then climb up

jump off

just keep swimming

sneak home

throw all those wet clothes

quick in the dryer

before mama gets home

she didn't want us out at Tar Creek

then someone tells us

about a big old swarm of snakes

come chasing out that junked fridge

no more going down that way

but you know

the water back then

don't think it harmed us
who knew?
wasn't all water kinda orange?

A scrappy can-do attitude kept Picher pride in what some called "Armageddon USA." Folks even had themselves a name— "chat rats." The chat piles (200 to 300 foot tall left over mine tailings) were playgrounds and Tar Creek and the mill ponds were swimming holes. I listened to a woman tell how her kids came back from the creek "all glittered." then wondering "what the creek all did." (personal communication Oma Potts)

Don't Call Me That
1996
Picher, Oklahoma

they call me dumb
retard
all the stuff I can't stand
makes me feel like punching
all those letters, numbers
bunched up in my brain
make me out
like some kind of fool
just don't get it
whatever it is

In the mid 1990s, health researchers found Picher had the highest level of juvenile lead poisoning in the United States. Kim Pace, a principal at the Picher-Cardin elementary school was interviewed "Our kids hit a brick wall. Their eyes skip and jump. It takes them 100 repetitions to learn a sound."

The Center for Disease Control declared in 2016, "no safe blood lead level has been identified. Even low levels of lead in blood have been shown to affect IQ, ability to pay attention and academic achievement. And most importantly, effects of lead exposure cannot be corrected." See Roosevelt

(2004) and Jim and Scott (2007) for stories of local residents dealing with high lead levels. Also Center for Disease control and Prevention—Lead website and Hu, Shine, and Wright (2007).

Tar Creek after the mines closed . In 1979, acid drainage from the mines started flooding into Tar Creek. Photo and permission given by Vaughn Wascovich.

Tar Creek Asks You to Take Another Look

April 2, 2016
Miami, Oklahoma

I see folks stare

right into my orange self

like I'm some kind of repugnant

aberration

wonder what they see

reflection, maybe?

always understood whatever we do

to each other

comes right back

maybe even ten fold

folks might think about that

when they let life

carry on just like nothing

they do or can do

makes much difference

maybe, just maybe when we really see each other

for what we are and could be

someone will wake up, call the alarm

find the power to do something

before it's too late

When I first saw Tar Creek, I thought vandals had purposely sprayed neon orange paint up and down the creek banks and in the water itself. Soon I learned how mining in the area was done by room-and pillar method. Rooms were carved out as the ore was removed. Pillars were left to keep the caves from caving in. When the mines closed in the late 1960s-early 70s water started filling these rooms, dissolving high concentrations of sulfide minerals and creating millions of gallons of acid water underground. In 1979, the acid drainage from the mines started flooding out into Tar Creek from numerous open mine shafts, natural springs, and boreholes. The surface water and some of the groundwater were instantly polluted. See the You Tube "Acid Mine Water at Tar Creek" and read the online University of Michigan's Tar Creek report for a more complete understanding of Tar Creek's history and toxicity.

Quapaw stories say they come from water. At one time, before taking a drink, a little water was spilled so Wah Kon Dah (God or Great Spirit) would know they were thankful for the gifts of mother earth. See Oklahoma Indian Affairs Commission (1977) for an understanding of Quapaw peoples' relationship with water.

Salamander Sings *Only the Lonely*
Early 2000s
Tri-State Mining District

I may be the one and only
salamander
in the whole Tri-State Mining District
kind of the canary
warning good or bad in the muck
of these wetlands and waters
they say I am an *indicator species*
can tell you about the health
of the ecosystem
we all got to share
better pay attention to me
maybe even join the campaign
Save the Salamanders
you never know
someday we might have to save you

Numerous studies, surveys and task forces have researched the Tar Creek Mining District, including the Kansas Biological Survey, the Harvard School of Public Health, the University of Michigan, the Center for Restoration of Ecosystems and Watersheds of the University of Oklahoma, the United States Environmental Protection

Agency. One of the investigators shared with me finding one salamander in an extensive study of the mining district. (personal communication Roger Kuhns)

Salamanders are considered "species indicators." Indicator species act as an early warning to monitoring biologists. Amphibians (including frogs, toads, and salamanders) are part of a group of organisms most sensitive to environmental change. For several years, there has been a severe decline in amphibians around the world, including the Tri-State Mining District. It's been said the salamander can serve the same function canaries did for coal miners. When canaries became sick or died due to being more sensitive to gases than humans, the miners knew something was wrong.

An Orange Rock in an Orange Creek

November 11, 2016
Tri-State Mining District

you think I'm toxic

scary

afraid to hold me

but in my cracks and fissures

I carry history

to touch me

is to know reality

a path

to hope of sorts

I keep an orange rock from Tar Creek with me— one also sits on my desk as I write this. I don't want to forget what we have done and what we are capable of doing.

Spook Light: ***A Mysterious Light of Unknown Origin***

Mid-1800s to present
Devil's Promenade, Oklahoma-Missouri State Line

funny thing about mystery
how we define
the real true story
how I flash my colors
bouncing down
the Oklahoma-Missouri state line
and how the stories
just keep creating
a miner
his illuminated cap light
searches for his kidnapped by Quapaws
lost child
and how at Devil's Promenade
some Indian maiden
is forever looking for her lost lover
even stranger how the government
concludes
a mysterious light of unknown origin
but this is what I know
we need mystery
in our lives
how truth is in the searching

the wonder
in the wandering
and I will keep my light bright

Along the Missouri-Oklahoma border, a road through a narrow canyon is known as the Devil's Promenade. Legend has it that the bouncing ball of light was seen back as far as 1836 when tribes were exiled to Indian Territory on their Trail of Tears It has had a variety of names—"The Hornet Spook Light," "Hollis Light," and the "Joplin Spook Light." In 1946, the United States Army Corps of Engineers did a study and described the phenomena as "a mysterious light of unknown origin." See Moran and Sceurman (2005) and Ferris (1995) for further "Spook Light" background and stories.

The aftermath of the 2008 Picher tornado.
Photo and permission by Vaughn Wascovich.

Dervish
May 10, 2008
Picher, Oklahoma

really
what else could I do
all those hot- cold currents
rising falling
echoes of trapped miners
pounding pow wow drums
sooners and Indians
pushing and pulling finally
I did what I thought could
only be right
like they say
clear the air
but first
pound my dervish self
pick up lay down
chat
houses
cars
a few human beings here and there
maybe now someone
will listen
really I'm not the wicked one

remember I left you
the chance to start over
and the funeral home
your choice

On May 10, 2008, an EF4 tornado slammed into Picher, Oklahoma. There were eight confirmed deaths and over 150 injuries. Twenty blocks of houses and businesses were destroyed or flattened. The funeral home remained standing. See You Tubes including "Picher tornado damage part 1& 2", and "Deadly tornado hits Picher, Oklahoma"

Fracking Mama
November 5, 2016
Cushing, Oklahoma

seems folks got a compulsion

to dig

but you start messing with me

I get pain

real bad pain

and when I hurt

I shake, rattle, and quake

then you, too

will know misery

but I've been thinking

how about we call a truce

take care of each other

remember Berry's scripture

do unto those downstream

as you would have those upstream do unto you

make those words true

and good

be family again

The website, Earthquake Track, keeps a daily tally of earthquakes throughout the world.

On November 18, 2016, they posted Oklahoma had

two earthquakes in the past 24 hours, twenty-five earthquakes in the past seven days, one hundred thirty-two earthquakes in the past thirty days, two thousand one hundred twenty-seven in the past 365 days.

From 1975 to 2008, Oklahoma averaged only one to three 3.0 magnitude or greater earthquakes annually. Oil and gas industry and fracking proponents argue that fracking does not cause earthquakes. Their literature reports that fracking process is not the issue but the burial of waste water. It is important, though, to understand that fracking cannot be done without producing large volumes of waste water.

Food and Water Watch, an environmental advocacy group has started to use the term "frackquakes." The Tri-State Mining District has good reason to fear even more environmental destruction. See Food and Water Watch's website for a variety of reports about fracking.

Remember what Wendell Berry said, "Do unto those downstream as you would have those upstream do unto you."

A 1967 Picher subsidence. Photo courtesy of Ed Keheley.

Picher in 2017—the buildings are torn down, the roads and chat piles remain.
Photo courtesy of the Tulsa World—Mike Simon.

Picher Blues
1915-present
Picher, Oklahoma

I was the end of the rainbow
where the Almighty
spilled his lead and zinc treasures
the jewel of the mining district
with my Billion Dollar Skyline
everyman's dream
of riches
now I'm all bought out
been called
the "Most Toxic Town in America"
even
"Armageddon USA"
I wait now
for green, the children
even the salamanders to start life over again

From pristine prairie to mandatory evacuation and buyout of an entire city, Picher's story is worthy of a biblical parable. The opening of rich mines in 1915 on the site of Picher led to overnight forced growth. On the 12th birthday anniversary of Picher, in 1927 a program extolled their city's 13,000 population that "sprang into existence as a necessity instead of the result of a blue-printed utopia of an idealist." A *Miami Record- Herald* mining editor called Picher a place "impossible to exaggerate." In 2006, an Army Corp of Engineers study showed 86% of its buildings undermined and ready to collapse. The state of Oklahoma and the Environmental Protection Agency agreed to a mandatory evacuation and buyout of the entire township. The municipality of Picher officially dissolved on November 26, 2013.

"Mealtime at a Quapaw Pow Wow" a 1940 lithograph by Charles Banks Wilson (1918-2013). Permission from the Gilcrease Museum.

"Quapaw Pow Wow" a 1941 lithograph by Charles Banks Wilson (1918-2013). The Quapaw Pow Wow is considered the oldest ongoing Pow Wow in the United States going back to 1872. Permission from the Gilcrease Museum.

The Echo Still Sings
July 4, 2009
Quapaw Pow Wow. Oklahoma

the old men tap the dirt

dance loose and light

hear fossil rhythms

the earth has held them so long

the young boys fancy dance

flashy and fast

medium war beat

ruffle

crow hop

fast beat

strength and stamina

someday the dirt will let us all go

then scudding leaves

free-falling clouds

maybe fox arteries

in the end

we become what we love

the echo still sings

The Quapaw Pow Wow is considered to be the oldest continuous annual pow wow in the country— the tribe celebrated their 147th in 2019, Back in 2009, I was especially drawn to the old men dancing and I stayed late into the night.

Quapaw folks are known for elaborate and generous pow wows and were especially well known during the heyday of royalties paid on lead and zinc deposits. Charles Banks Wilson, Oklahoma's most famous artist (and married to Edna McKibben, a Quapaw woman) lithographed and described a pow wow mealtime in 1940..." the food—dried corn, wild grape dumplings, squaw bread, pecan butter, wild meat, and beef- probably brought as many Indians to their festivities as did the colorful dances." See Hunt & Wilson (1989), Nieberding (1976), and Wilson (1964) for Quapaw pow wow particulars.

One of the old men I watched dance that night died two months later. The echo still
sings...

A Glossary of Names, Places, Terms

A

B

Baxter Springs—a city in Cherokee County, Kansas on the border of Indian Territory— Oklahoma and Kansas. The original spring which made the town famous for its healing waters ceased to flow with the advent of lead and zinc mining.

Blue Card Union—a company union formed in response to striking mining workers in 1935. The pick handle was the badge of Blue Card members.

C

Chat—fine gravel waste left over after lead and zinc ore production. Chat contains toxic levels of lead, zinc, and cadmium

Chat Piles—scattered piles of chat, some as tall as 200 feet and four football fields in size. When the Tri-State Mining District stopped operating, some 175 million tons of chat were left behind.

Chat Rats—Picher residents, also been called "lead heads."

Commerce-—a city in Ottawa County in northeastern Oklahoma on land that was originally the Quapaw Indian Agency. A mining camp named Hattonville was formed in

1906. The Commerce Mining and Royalty Company bought out the camp and the town's name changed to Commerce.

Cookson Hills—is an area in eastern Oklahoma originally part of the Cherokee Nation. Outlaws and fugitives, including Pretty Boy Floyd, The Dalton Gang, Belle Starr, and the James Brothers often chose the hills as their sanctuary from the law.

D

Douthat— a ghost town in Ottawa County in northeastern Oklahoma and part of the Tar Creek Superfund site. It was the site of giant lead and zinc mills.

Devil's Promenade—an area on the border between southwestern Missouri and northeastern Oklahoma where a single ball of light or a tight grouping of lights is said to appear regularly at night. The light is known as the **Spook Light.**

E

Eagle-Picher Mining Company—was the most prominent, largest, and longest lived enterprise in the Tri-State Mining District. Its mining operations shut down in 1967.

F

Fracking—is a method in which water, chemicals, and sand are injected at high pressure into a well drilled in a shale formation to break up the rock and release gas and oil.

G

H

Hiawatha-Canton Insane Asylum for Indians—a federal institution in Canton, South Dakota in operation from the early 1900s to the 1930s when it was closed following investigations of horrendous conditions and practices. It has been called a dumping ground for inconvenient Indians.

Incompetent Restricted Indian—is a legal term referring to an Indian declared incompetent by the federal government—the Bureau of Indian Affairs and in the Indian Territory-Oklahoma lead-zinc district a guardian would have been appointed by the county commissioners of Ottawa County. His/her allotted land would be under restrictions from selling or leasing.

I

Indian Territory—IT—was land within the United States of America reserved for the forced re-settlement of Native Americans. The Organic Act of 1890 reduced IT to the lands occupied by the Five Civilized Tribes and the Tribes of the Quapaw Indian Agency. The western part of Indian Territory became Oklahoma Territory. In November 1907, Indian Territory and Oklahoma Territory became a single state—Oklahoma.

Indicator Species—is an organism whose presence, absence or abundance reflects a specific environmental condition. It can signal a change in the biological condition of a particular ecosystem—can be used as an "early warning system."

J

K

L

LEAD agency—Local Environmental Action Demanded —is a non-profit corporation organized in 1997. It works out of Miami, Oklahoma with goals to educate the community on environmental concerns, take action to counter environmental hazards, conduct environmental workshops and seminars, and partner with other environmental organizations.

M

Miners' Con—see **Silicosis**

Modoc—is the smallest federally recognized Indian tribe in Oklahoma. Their original home was on the Oregon-California border where the Modoc War was fought in 1873. Their leaders were hanged or sent to Alcatraz and the rest of the tribe were exiled to the Quapaw Agency, Indian Territory.

Mother Road–Route 66–Main Street of America–Will Rogers Highway—was one of the original highways within the U.S. Highway System. John Steinbeck in *The Grapes of Wrath* wrote, "66 is the mother road, the road of flight." The two halves of Route 66 met in the town of **Quapaw** leading some to refer to the town as "where east meets west." The highway stretched from Chicago to Santa Monica, California.

MVP—most valuable player

N

Nez Perce—Nimiipuu—an American Indian tribe exiled to Indian Territory following a flight of over 1,000 miles from their Oregon-Idaho homeland almost to the Canadian border. The last battle was in the Bear Paw Mountains of Montana. As prisoners of war, the tribe was originally sent to Fort Leavenworth and then on to the Quapaw Agency in Indian Territory.

O

Ottawa County—a county in northeastern Oklahoma. **Miami** is the county seat and was the federal court town of the Quapaw Agency. In a 1902 promotional booklet, the writer congratulated Miami "because there is not a colored person living in the town and few foreigners or full blooded Indians, the population being mostly American born." Boasting their water supply, the writer continued to say "If Miami was not otherwise supplied with water for all purposes the Neosho river and Tar Creek would furnish a never failing supply of the purest and clearest."

P

Peoria—is a town in Ottawa County, Oklahoma. It was named for the Peoria tribe who were moved into Indian Territory in the 19th century. Commercial lead and zinc mining in the area was first carried out in 1891.

Peyote—is the communion of the Native American Church said to allow participants to experience God directly. Peyote contains the hallucinogenic drug mescaline and is not addictive.

Picher—now a ghost town in northeastern Oklahoma was, at one time, one of the most productive lead and zinc mining areas in the world. In 1915, drilling operations led to the opening of rich mines on the site of Picher and what was originally Quapaw Indian land. The area was declared the Tar Creek Superfund site in 1981 when the United States Environmental Protection Agency called it the most toxic place in America. It was home to 14,000 abandoned mine shafts, 70 million tons of mine tailings, and 36 million tons of mill sand and sludge. Eventually, the town was deemed too toxic to clean up and a federal buyout program paid people to leave. The city ceased operations as a municipality on September 1, 2009.

Q

Quapaw—is a town in northeastern Oklahoma and part of the Tri-State Mining District. Prior to the discovery of lead and zinc, Quapaw was known as the "Hay Capital of the World." The town is "where east meets west" on **Route 66**.

Quapaw Indian Agency—was territory spanning 220,000 acres that included parts of what are now Ottawa and Delaware counties in Oklahoma. It was established in the late 1830s. The primary tribes of the agency were Eastern Shawnee, Miami, Modoc, Ottawa, Peoria, Quapaw, Seneca and Cayuga, and Wyandotte. As a federal Indian agency, it operated to manage the relationship between the federal government and the various tribes.

Quapaw Tribe of Indians—O-Gah-Pah—Ugaxpa—Ogaxpa—Okaxpa—Arkansea—- Downstream People—is a federally recognized tribe based in Ottawa County, Oklahoma. In 1834, the Quapaw were removed from the Mississippi valley areas to the northeast corner of what was then Indian Territory, now Oklahoma. The Quapaw tribal jurisdictional area includes what became the Tar Creek Superfund Site. A high percentage of Quapaw and other residents of Ottawa County suffer from lead poisoning and other ill effects from the contamination of ground and water.

R

S

Seneca Indian School—originally founded by Quakers in 1868, the school was first known as the Wyandotte Mission. They called it "a Light in the Wilderness" or a "Godly Experiment in the Wilderness." By 1910, the school was under government direction. Seneca Indian School was the oldest Indian boarding school in continuous operation (112 years) when it closed in 1980. It was located near Wyandotte, Indian Territory-Oklahoma.

Silicosis—also known as **Miners' Con** or **Miners' Consumption**—a dust induced respiratory infection predisposing workers to tuberculosis. In crowded, poorly-housed mining camps, family members often also became infected with tuberculosis.

St Mary's of the Quapaw School—was a school established by the Catholic church in 1894 and closed in 1927 near what is now Quapaw, Oklahoma. Some believed the school was closed in reaction to anti-Catholic Ku Klux Klan activity.

Superfund—is a United States federal government program designed to fund the cleanup of sites contaminated with hazardous substances and pollutants. It was established by the Comprehensive Environmental Response Compensation and Liability Act of 1980. Its goal is to protect public health and the environment.

Subsidence—is the sinking of the earth's surface in response to geologic or man-induced causes. A **sinkhole** is a depression in the ground that has no natural external surface drainage.

T

Tar Creek—is a creek in northeastern Oklahoma—southeastern Kansas and a 40 square- mile site contaminated with remains from what was once one of the largest lead and zinc mining operations in the world.

Tri-State Mining District—located in southeast Kansas, southwest Missouri, and northeast Oklahoma where lead and zinc mining took place.

U

V

W

Wyandotte Nation of Oklahoma—a federally recognized tribe located in northeastern Oklahoma. They were the last tribe east of the Mississippi to be removed to Indian Territory.

Wyandotte, Oklahoma—a town in Ottawa County, Oklahoma and headquarters of the Wyandotte Nation of Oklahoma for which the town was named. The Society of Friends (Quakers) started a mission there in 1869.

X

Y

Z

References

500NATIONS.COM. (2017).
A website providing information re: Indian gaming/gambling throughout the United States, including Oklahoma.

Anderson, D., Yadon, L. J., & Smith, R. B. (2007). Bonnie and Clyde and Them. In *100 Oklahoma outlaws, gangsters, and lawmen, 1839-1939* (pp. 103-117). Gretna, LA: Pelican Pub. Co.
Information re: their Commerce, Oklahoma shooting and kidnapping episode.

Baird, W. D. (1975). *The Quapaw People*. Phoenix, AZ: Indian Tribal Series.
A short history of the Quapaw tribe.

Baird, W. D. (1980). *The Quapaw Indians: A history of the Downstream People*. Norman: University of Oklahoma Press.
An extensive well-researched history of the Quapaw tribe.

Barringer, F. (2004, April 12). Despite Cleanup at Mine, Dust and Fear Linger. *The New York Times*.
A graphic description of the Tar Creek Mining District-Superfund site.

Brown, C. E. (1975). *Program 12th Birthday Anniversary City of Picher, Oklahoma*. Miami, OK: Dixon Printing & Stationary Co.
Originally published in 1927 as a promotional booklet for Picher.

Channing L. Bete Co. (1996). *Your Child's Lead Level--What does it mean?* South Deerfield, MA: Channing L. Bete Co.
An old but good basic overview of lead.

Channing L. Bete Co. (1996). *What everyone should know about lead poisoning*. South Deerfield, MA: Channing L. Bete.

An old but good basic overview of lead poisoning.

Cherokee Nation, Miami High School (Miami, Okla.). Tar Creek Project, & Miami High School (Miami, Okla.). Cherokee Volunteer Society. (1999). *Tar Creek anthology, "the legacy": Poems, songs, essays and research papers*. Tahlequah, OK: Tahlequah Daily Press.
An inspiring anthology of writings by young people affected by the Tar Creek environmental disaster.

Cook, F. L. (2011). *Pictures from the mining era: Cardin-Picher, OK area*. Wyandotte, OK: Gregarth Pub. Co.
A large collection of old Tri-State Mining District photographs.

Cook, F. M., Burns, N. L., & United States. (2012). *Nannie Lee Burns Indian Pioneer History interviews of Ottawa County, Oklahoma, 1937-1938*. Wyandotte, OK: Gregath Co.
Depression era interviews of Ottawa County residents.

Cottrell, S. (1995). *Civil War in the Indian Territory*. Gretna, LA: Pelican Pub. Co.
Little known history of Indian Territory during the Civil War.

Craig, G. S. (1976). *Picher, Oklahoma The Lead and Zinc Boom Town that would not die*. Picher, OK: Picher Bicentennial Boosters Committee.
Part of a promotional booklet.

Cray, E. (2004). *Ramblin' Man The Life and Times of Woody Guthrie*. New York, NY: W.W. Norton & Company.
Good background on how Pretty Boy Floyd became both legend and song.

Davidson, L. S. (1939). *South of Joplin: Story of a Tri-State Diggin's*. New York: W.W. Norton.
A muck racking novel depicting life in the Tar Creek Mining District during the Depression and strike.

Debo, A., & Writers' Program of the Work Projects Administration in the State of Oklahoma. (1986). *The*

WPA guide to 1930s Oklahoma. Lawrence, Kan: University Press of Kansas.
The federal government hired writers during the 1930s Depression to write state guides.

Denworth, L. (2008). *Toxic truth: A scientist, a doctor, and the battle over lead*. Boston: Beacon Press.
A helpful guide to understanding the toxicity of lead.

Dick, L. (Director). (1940). *Men and Dust* [Motion picture]. USA.
A classic film documentary about the Tar Creek Mining District and its people.

Eagle-Picher 150 Years of Quality. (1993). Retrieved from Eagle-Picher Industries, Inc. website:
A promotional booklet put out by the Tar Creek Mining District's biggest employer.

Earthquaketrack. (2017). Retrieved from http://earthquaketrack.com
An ongoing record of earthquake activity throughout the world.

EPA Fact Sheet Tribal Leadership, Historic Preservation and Green Remediation. (2015). United States Environmental Protection Agency.
Documentation of the work and coordination between the Quapaw Tribe and the US Environmental Protection Agency in the cleanup of the Catholic 40, part of the Tar Creek Superfund Site.

Farris, D. A. (1995). *Mysterious Oklahoma: Eerie true tales from the Sooner State*. Edmond, OK: Little Bruce.
Information about the Spook Light.

Fischer, L. R., & Rampp, L. C. (1968). *Quantrill's Civil War operations in Indian Territory*. Oklahoma City: Oklahoma Historical Society.
Information about Quantrill's Raiders, also known as Bushwhackers (pro-Confederate partisans) during the Civil War.

Food & Water Watch (Organization). (2016). *The Urgent Case for a Ban on Fracking*. Washington, DC: Food

& Water Watch.
An important report on fracking
Fugate, F. L., & Fugate, R. B. (1991). *Roadside history of Oklahoma*. Missoula, MT: Mountain Press Pub. Co.
Little known facts about Oklahoma.
Gibson, A. M. (1972). *Wilderness bonanza: The Tri-State District of Missouri, Kansas, and Oklahoma.* Norman: University of Oklahoma Press.
A wealth of information about mining in the Tri-State district.
Grass, A. (2016). *Fracking Causes Earthquakes. Period.* Food & Water Watch.
Well-documented information re: the correlation between fracking and earthquakes.
Hoover, C. (1926). *Dixon's Oklahoma-Kansas Mining Directory* (reprint). Miami, OK: O.T. Dixon Printing and Stationary Co.
An astonishing number of mines are listed in this booklet originally published in 1926.
Hu, H., Shine, J., & Wright, R. O. (2007). The challenge posed to children's health by mixtures of toxic waste: The Tar Creek Superfund Site as a case-study. *Pediatric Clinics of North America*. Retrieved from https://www.ncbi.nim.nih.gov
An extensive study concerning lead and children in the Tar Creek Superfund Site.
Hunt, D. C., & Wilson, C. B. (1989). *The lithographs of Charles Banks Wilson*. Norman, OK: University of Oklahoma Press.
A collection of lithographs done by one of Oklahoma's most famous artists. Many depict Tar Creek, miners, Quapaw Indians.
Jim, R., Scott, M., & LEAD Agency. (2007). *Making a difference at the Tar Creek Superfund site: Community efforts to reduce risk*. Vinita, OK: LEAD Agency, Inc.
A helpful guide to understanding grassroots efforts to

deal with Tar Creek's devastation.

Johnson, L. G. (2008). *Tar Creek: A history of the Quapaw Indians, the world's largest lead and zinc discovery, and the Tar Creek Superfund site*. Mustang, OK: Tate Pub. & Enterprises.
A broad overview of the history of mining, Quapaw Indians, environmental issues.

Joinson, C. (2016). *Vanished in Hiawatha: The story of the Canton Asylum for Insane Indians*. Lincoln, NE: University of Nebraska Press.
A helpful reference for understanding the federal government's only institution for "insane" Indians.

Kreman, C. (2014, June). *Tribal-Led remedial Action at the Tar Creek Superfund Site*
[PowerPoint].
Power Point presentation by the Quapaw Tribe Environmental Office to the American Society of Mining & Reclamation national meeting in Oklahoma City providing details about the history and the current clean up work at the Tar Creek Superfund Site.

LEAD Agency. (2015). Miami, OK.
Swimmable Fishable Drinkable 17th Tar Creek National Environmental Conference booklet put out for the LEAD agency annual conference.

Leahy, T. (2009). *They called it madness: The Canton Asylum for Insane Indians,1899-1934*. Baltimore, MD: PublishAmerica.
Further information about the Canton-Hiawatha Insane Asylum for Indians.

Leavy, J. (2010). October 5, 1951 When Fates Converge. In *The last boy: Mickey Mantle and the end of America's childhood* (p. 20). New York, NY: Harper Collins Publishers.
Chronicles the life of Mickey Mantle- good information about his early life in the mining district.

Luza, Kenneth V, & Keheley, W. E. (2006). *Field trip guide*

to the Tar Creek superfund site, Picher, Oklahoma: Oklahoma Section, American Institute of Professional Geologists, Annual State Meeting, April 28-29, 2006, Shangri La Resort, Grand Lake, Oklahoma. Norman, Okla.: Oklahoma Geological Survey.
An extensive well-researched geological and historical study with photographs and graphs of the Tar Creek Superfund Site and the Tri-State Mining District.

Martinez, L. R. (2007). *Hard rock legacy: Memories and history of Picher Oklahoma*. Picher, OK: Sunset Pub. Co.
A homegrown book with many photographs and stories of Picher.

Martinez, L. R. (2008). *Picher: 90 years of memories*. Picher, OK: Lynda Ramsey Martinez.
More photographs and stories of Picher.

Martinez, L. R., Ray, O. H., & Koutz, M. R. (n.d.). *From Whistle to Whistle Lead and Zinc Mining Terms Picher, Oklahoma*. self published.
Includes a glossary of locally used mining terms.

Meyer, R. E. (1983, February 2). The Tar Creek Time Bomb. *Los Angeles Times*.
One of the first nation-wide articles to detail Tar Creek's story.

Minerals, Mineral Industries and Reclamation. (1977). In J. W. Morris (Ed.), *Geography of Oklahoma* (pp. 93-111). Oklahoma City, OK: Oklahoma Historical Society.
A helpful reference to understand the scientific background of mining in Oklahoma.

Morris, J. W. (1982). Lead and Zinc. In *Drill bits, picks, and shovels: A history of mineral resources in Oklahoma* (pp. 112-131). Oklahoma City, OK: Oklahoma Historical Society.
A helpful reference to understanding of Oklahoma's mineral resources.

Myers, M. (Director). (n.d.). *Tar Creek* [Motion picture].
A powerful documentary depicting Tar Creek and its history.

Nerburn, K. (2013). *The girl who sang to the buffalo: A child, an elder, and the light from an ancient sky.*
An account of an old man and a writer searching for the old man's little sister's story. She had been sent to Hiawatha-Canton Insane Asylum for Indians.

Nieberding, V. (1976). *The Quapaws: Those who went downstream* (3rd ed.). Wyandotte, OK: The Gregath Publishing Company.
A local writer's account of the Quapaw Indians.

Nieberding, V. (1983). *The History of Ottawa County.* Marceline, MO: Walsworth Publishing Company.
An extensive book of Ottawa County, Oklahoma, its people, history, land.

NRE 492 Group 5. (n.d.). The Results of Mining at Tar Creek. Retrieved from http://umich.edu/-snre492/cases-0304/TarCreek/TarCreek-case-study.htm
University of Michigan's Tar Creek study.

Odell, R. M. (1972). *A Pen Picture of Miami, Indian Territory, and Tributary Lands.* Miami, OK: Dixon Printing and Stationery Company.
Originally published November 1902 as a promotional guide to Miami and the Tributary Lands,

Oklahoma Indian Affairs Commission. (1977). *Ogaxpa.* Oklahoma City, OK: The Commission.
A collection of first person elderly Quapaw stories.

O'Neal, L. D. (2006). *Nez Perce Exile: The Struggle for Freedom 1877-1885.* Wallowa, OR: Bear Creek Press.
An overview of the Nez Perce exile.

Osborne, D. (2017). *The Coming.* New York, NY: Bloomsbury.
A novel representation of Daytime Smoke, also known as Halahtookit, the son William Clark and a

Nez Perce woman.
Parker, M. (2006). *Killed & injured at or in the mines, Ottawa Co., Okla*. Welch, OK: Margaret Parker.
A two volume compilation of newspaper articles concerning mining deaths and injuries in Ottawa County.
Pearson, J. D. (2008). Life in the Eeikish Pah, the Hot Place. In *The Nez Perces in the Indian territory: Nimiipuu survival* (pp. 114-145). Norman, OK: University of Oklahoma Press.
A valuable resource for understanding the Nez Perce experience at the Quapaw Agency in Indian Territory.
Robertson, D. (2010). Picher. In *Hard as the rock itself: Place and identity in the American mining town* (pp. 121-183). Boulder, CO: University Press of Colorado.
An extensive overview of Picher mining and history.
Roosevelt, M. (2004, 26). The Tragedy of Tar Creek. *Time*, 42-47.
Time Magazine provided Tar Creek with national coverage.
Rosner, D., & Markowitz, G. E. (2006). *Deadly dust: Silicosis and the on-going struggle to protect workers' health*. Ann Arbor: University of Michigan Press.
An especially informative book providing a clear understanding and history of mining issues, including the Tri-State Mining district, and public health.
Ross, M. (1939). Rise and Fall of a Union, Local Opinion209. In *Death of a Yale man* (pp. 189- 209). New York, NY: Farrar & Rinehart, Inc.
The chapter cited concerns the 1935 miners' labor strike.
Ruth, K., University of Oklahoma, & Writers' Program of the Work Projects Administration in the State of Oklahoma. (1957). *Oklahoma: A guide to the Sooner State*. Norman, OK: University of Oklahoma Press.
An update of the Oklahoma state writers' guide.
Sceurman, M., Moran, M., & Lake, M. (2008). Unexplained

phenomena. In *Weird U.S: The odyssey continues: your travel guide to America's local legends and best kept secrets* (pp. 100-101). New York, NY: Sterling.
A variety of Spook Light legends are included.

Snyder, G. (2003). *A Legacy for Luke* (pp. 173-181). Cassville, MO: Glen Haven Publisher.
A novel depicting mining life in the Tri-State Mining District.

Stewart, T., & Fields, A. (2016). *Picher, Oklahoma: Catastrophe, memory, and trauma*. Norman: University of Oklahoma.
An almost coffee table book detailing Picher's story with photographs and essays

Suggs, G. G. (1986). *Union busting in the Tri-State: The Oklahoma, Kansas, and Missouri metal workers' strike of 1935*. Norman: University of Oklahoma Press.
The most extensive writing on the 1935 labor strike.

United States Congress House Committee on Indian Affairs. (1919). *Indian Appropriation Bill Supplemental Hearing Before Subcommittee of the Committee of Indian Affairs of the House of Representatives*. Retrieved from a primary resource that helps understanding of the federal government's involvement with the Quapaw Tribe

United States, & Hitchcock, E. A. (1907). *Regulations to be observed in the leasing for mining purposes of allotted lands of incompetent Indians of the Quapaw Agency, Indian Territory: Prescribed by the Secretary of the Interior, January 24, 1907, for the purpose of carrying into effect the provisions of the act of Congress approved June 7, 1897 (30 Stat. L., 72)*. Washington DC: Government Printing Office.
A primary resource outlining the regulations prescribed to govern the leasing of Quapaw Agency allottees.

Wallis, M. (1993). Route 66: the Mother Road Pretty Boy. In

Way down yonder in the Indian nation: Writing from America's heartland (pp. 29-57 156-175). New York: St. Martin's Press.
Local color stories of Oklahoma.

Wannenmacher, P. (Director). (2013). *The Ozark Uplift The Story of Tri-State Mining* [Motion picture]. Missouri USA: Ozarks Public Television.
A film documentary that includes interviews with old miners.

Warde, M. J. (2013). *When the wolf came: The civil war and the Indian territory.* Fayetteville, AK: The University of Arkansas Press.
Helpful reference for understanding the Civil War in Indian Territory.

Watt, S. (2002). *Lead.* New York, NY: Benchmark Books.
An old but helpful book for facts about the mineral lead.

Wikipedia. (n.d.). Picher, Oklahoma.
Basic overall information re: Picher Oklahoma

Wills, D. (1974). How the Picher District Really Came to Be. *True West*, *21*(6), 24-25 62-63.
Information about some of the first commercial mining in the Picher Mining District.

Wilson, C. B. (1964). *Indians of Eastern Oklahoma.* Afton, OK: Buffalo Publishing Company.
Helpful information, photographs, and art work about eastern Oklahoma's small tribes written by one of Oklahoma's most famous artists. His roots were in eastern Oklahoma.

Wilson, C. B., Ramer, R., Haralson, C., Klein, C., Roblin, K., Finegan, K., Gilcrease Museum. (2007). *Charles Banks Wilson.* Tulsa, OK: Gilcrease Museum.
A collection of Charles Banks Wilson's work.

Wilson, S. (1989). Tales of the Indian Nations. In *Oklahoma treasures and treasure tales*
(pp. 249-257). Norman, OK: University of Oklahoma Press.

Origins of Peoria mining.
Wright, M. H. (1957). *A guide to the Indian tribes of Oklahoma*. Norman, OK: Oklahoma City.
Short histories of Oklahoma tribes.
Zitkala-Š̈ a, Fabens, C. H., & Sniffen, M. K. (1924). *Oklahoma's poor rich Indians: An orgy of graft and exploitation of the Five Civilized Tribes, legalized robbery : a report.*
Philadelphia, PA: Office of the Indian Rights Association.
A book that influenced both public feeling and government action re: Indian treatment in Oklahoma. A very short section mentions Robert Thompson, a Quapaw Indian sent to Hiawatha Insane Asylum for Indians.

Epilogue

Tar Creek 2021

As *Once Upon a Tar Creek: Mining for Voices* came to completion further destruction of our environment is evident everywhere and questions of cultural appropriation have become bigger and bigger. The outrage I felt when I first saw Tar Creek has only grown but with some glimmers of hope.

Flint, Michigan has experienced and continues to suffer from lead infested water. The Clean Water Act has regulated discharges of contaminants since 1972. In April 2020, our previous administration finalized a new definition of what marshes, wetlands, and streams qualify for protection. Bears Ears has lost protection under the guise of local control. Every day, we lose more and more conservation of our most basic natural resources.

Black Lives Matter and native peoples' resistance to what some call environmental genocide have raised questions on who speaks for whom. Inevitably, the big questions outnumber the answers. Writing *Tar Creek* has involved hearing then speaking voices not always my own. I have been blessed to have encountered many people willing to trust me with their stories. Grace Goodeagle, Quapaw tribal elder and first woman elected chairman of her tribe "Quapaw fact-checked" this manuscript. Arnold Richardson and his daughter, Oma Potts (now both deceased) shared many painful memories. I have become even more convinced as Barry Lopez writes, "sometimes a person needs a story more than food to stay alive."

Writing this book, I hoped to communicate my fear for and love of our earth home. The more we know our connections to each other, the more chance we will live our lives respectfully and responsibly. I still have reason for hope, although with many reservations. In 2017, LEAD (Local Environmental Action Demanded) Agency celebrated twenty years of advocating for Tar Creek and they continue to be going strong. Rebecca Jim, their executive director, has officially become the Waterkeeper Alliance's Tar Creekkeeper. Native and more and more mainstream Americans have become "Water-Protector-Warriors." The controversial Keystone XL tar sands pipeline was dealt a major setback in April 2020 after a judge revoked a key permit issued by the US Army Corp of Engineers that did not properly assess the impact on endangered species and now in 2021, the project is now in question again. In 2012, for the first time in the Environmental Protection Agency's Superfund program, a tribe has taken over the cleanup of a contaminated site. The Quapaw Tribe signed an agreement with the EPA to protect and preserve the "Catholic 40"—a site where many tribal members attended boarding school and church. The tribe is now extending their work on sites throughout the former Tri-State Mining District. In April 2017, the March for Science saw thousands marching in support of our environment. The United States, as of 2021, is again a member of the Paris Climate Agreement.

Our very existence depends on water. Our bodies are 62% water. Maybe when we recognize our absolute connection to water, we will stand passionate in its protection. Yes, water is life and we are all both up and downstream.

Acknowledgments

If it takes a village to raise a child, it takes a cast of hundreds to complete a book.

The Cantwell clan have always been incredible storytellers. I thank them for always remembering to remember.

Gratitude to my parents, Al and Wanda Cantwell Hurtt, who provided an endless supply of books, pens, pencils, and paper, read to me John Muir stories, silly poems, and the Sunday comics, and most of all, a forever sense of wonder. My father told me before he died that when we read the comics together, I called question marks "wonder marks." I am thankful they gave me their own sense of awe and wonder. My brother, Jim Hurtt, is proof positive that love is so much more than blood. Your humor sustains me. Our parents are gone now, but when we were cleaning out their house, we found a copy of an old Christmas letter she had sent out years ago: "Maryann and I met each other in Oklahoma this year. She wants to write about the miners and the Indians." I am sure she would be pleased.

Thank you to my writing pals for encouragement and listening to endless Tar Creek stories: the Grand Avenue Poetry Collective (the Posse), Sandy Rockhill (title person extraordinaire), Kay Sanders, Mandi Issacson, and the Oshkosh gang, the Mead Library once a month poetry fix, and fellow Maywood Electric Squirrels—Marilyn Zelke-Windau and Georgia Ressmeyer. Thanks also to the Prague Summer Program, Fishtrap, Breadloaf-Orion, the Clearing, the Scissortail Writing Festival, and Ghost Ranch for amazing writing experiences and scholarship support.

Where would a woman be without girlfriends? Karen, Ann Janette, Kristi, Judy, Cindy, Andrea, Terri, Sara, Amber, Virginia, Susan, Beth... you have stayed by my side through blizzards, locked doors, midnight calls, broken and joyful hearts and proved over and over again our sisterhood.

Both Arnold Richardson and his daughter, Oma Potts shared first hand Indian Territory- Oklahoma stories, fed me, and took me on back roads and back stories. Both are gone now but their stories carry on. Retha Epperson, Arnold's granddaughter-Oma's daughter helped give context to the stories.

The folks at the LEAD (Local Environmental Action Demanded) Agency in Miami, Oklahoma, especially Tar Creek keeper Rebecca Jim and Grand River keeper Earl Hatley, deserve undying respect for their knowledge, commitment, and passion in making our earth healthy again. Thank you for inviting me to read at the Tar Creek Environmental Conference. Thank you also to Ed Keheley for your wealth of everything Tar Creek knowledge. Thanks also to the late Jackie and Harlan Brewer for sharing their home, garden, family competency papers, and memories. I wish they were alive to see the results of their kindness.

Appreciation to Carole White, Kay Sanders, Julie Stodolka, Ann Lehman, Susan Huebner, and Sylvia Cavanaugh for reviewing and proofreading this manuscript in its many versions.

Thank you, Grace Goodeagle for "Quapaw fact checking" *Once Upon a Tar Creek* and believing in the power of stories and the need to remember the stories.

Gratitude also to Turning Plow Press editor Paul Bowers, Rowan Kehn, and Oklahoma poet-teacher Ken Hada for making the stars align. *Once Upon a Tar Creek: Mining for Voices* found a home through their kindness and work.

Last of all, I thank Steven R. Manthey, husband extraordinaire, for all the computer assist, trips to the airport, scrabble games, tandem rides, ice fishing, Friday night pizza dates, a household full of Tar Creek research, and responding with good humor to "now where is Maryann going?" I will be forever grateful for you having my back.

A version of "And the Echo Still Sings" (as "The Old Man at the Pow Wow") was originally published in *Verse Wisconsin* Spring 2010 and in *River* by Maryann Hurtt, Aldrich Press, 2016. "Fracking Mama" was published as part of LEAD Agency's National Environmental Tar Creek Conference. "My Repugnant Tar Creek Orange Self" was published in *The Water Poems*, Water's Edge Press, 2017

About the Author

Growing up in Livermoore, California, Maryann Hurtt aspired to be a rodeo clown. By the time she was in 6th grade in Arlington, Virginia, she recorded in her diary that she wanted to be a "storyteller (a good one)." She now lives in Wisconsin's Kettle Moraine and is retired after thirty years working as a hospice RN. She spends her time hiking, reading, biking, and writing.

Hurtt's poetry has been published in a variety of print journals, online, and anthologies including *Verse Wisconsin, Poetry Hall, Ariel, Stoneboat, Verse-Virtual, Wisconsin People & Ideas, Blue Heron Review, Halfway Down the Stairs, Snapdragon, Cancer Poetry Project II*, and others. Her chapbook, *River*, came out in 2016. She also co-authored with Cynthia Frozena, *Hospice Care Planning: An Interdisciplinary Guide*.

She is passionate about environmental issues and has shared her Tar Creek poetry in Wisconsin, Oregon, Oklahoma, Montana, Vermont, New Mexico, Arizona, and Washington, D.C. She is forever grateful to Water and Water Protectors everywhere.